Dark ≠ Hidden

Sandeep Chavan

Published by SANDEEP CHAVAN, 2025.

While every precaution has been taken in the preparation of this book, the publisher assumes no responsibility for errors or omissions, or for damages resulting from the use of the information contained herein.

DARK ≠ HIDDEN

First edition. July 13, 2025.

ISBN: 979-8231657940

Written by Sandeep Chavan.

Table of Contents

DedicationTo **Fritz Zwicky**—who, in 1933, first pointed toward something "missing" in the Coma Cluster,not knowing that his observation would echo through generations.

This book is not written to contradict you—but to complete the question you bravely asked.

You saw the consequence.The world invented a substance.We return now—not to disprove the darkness you glimpsed,but to finally read it... as a ripple.

With respect,**Sandeep Chavan**

UED in One Breath

The universe manifests—so there is existence.But existence is not made of objects; it is made of ripples.Before the observer, it ripples.When the ripple resolves into consequence, it becomes matter, life, and energy.This resolution—this alignment—is what we call physical reality.

We measure, manipulate, and build knowledge only from resolved consequences.But true understanding is impossible without two forgotten presences:the **Structured Vacuum Energy (SVE)** that gives rise to the ripple,and the **Observer** who tunes into its alignment.

Without these, science remains clever—but blind.With them, we realign—not just our models—but our understanding of reality itself.

— *Sandeep Chavan*

Disclaimer

This book presents a conceptual framework known as **Universal Energy Dynamics (UED)**—a new interpretive model that re-examines fundamental scientific assumptions, including those in physics, cosmology, and observational methodologies. While the author has made every effort to ensure logical coherence and internal consistency, this work does not claim to be a replacement for rigorous peer-reviewed scientific consensus unless or until it is validated through empirical experimentation.

Any reinterpretations of established theories such as Newtonian mechanics, Einstein's relativity, or quantum field theory are offered as intellectual contributions intended to provoke inquiry, inspire new models, and encourage deeper scientific and philosophical engagement. Readers are advised to approach this work as **an alternative viewpoint grounded in consequence-based reasoning**, not as a definitive or authoritative scientific doctrine.

The discussions, models, equations, and proposals herein should not be construed as academic advice, formal educational content, or claims of immediate technological viability. They are **thought experiments**, reformulations, and visionary constructs intended to push the boundaries of inquiry.

The author disclaims any liability arising directly or indirectly from the use, application, or misinterpretation of the ideas and concepts discussed in this book.

In short: This is a contribution to the *future of thinking*, not a final verdict on the *facts of today*.

Preface

We live in a universe that, by all appearances, works—but we no longer know why.

For over a century, physics has been in a silent crisis—patched together with equations that function but don't explain, with models that predict but don't align, and with constants that behave more like placeholders than principles. Nowhere is this more apparent than in the way we speak of the "dark": dark matter, dark energy, dark flow—concepts born not of clarity, but of confusion measured precisely.

This book does not chase shadows. It questions the flashlight.

What follows is not an argument against science—it is a call to complete it. *Universal Energy Dynamics (UED)* was not born from dissatisfaction, but from consequence. It arose because the more we tried to force the universe into boxes built by force and mass, the more it refused to fit. So instead of asking "what else is out there," UED asks, *"what are we overlooking right here?"*

At the core of UED lies a simple pivot: that energy behaves not as a push or pull, but as a ripple seeking realignment in a structured, memory-rich vacuum field. What we call mass is a ripple configuration. What we perceive as gravity is field tension. What we label dark is simply unaligned understanding.

This book takes no pleasure in dismissing dark matter or dark energy. Instead, it lovingly acknowledges the necessity of their invention in a model that had run out of answers—but not out of questions. UED is that next question.

You will find no hidden particles here, no new forces, no mysterious fifth elements. You will find consequences—measurable, traceable, testable. You will find predictions—not because they are clever, but because they are aligned. And most importantly, you will find silence—the structured stillness before motion—finally given its place in the conversation.

If you are a physicist, this book will challenge your habits. If you are a seeker, it will align with your intuition. And if you are simply curious, it will give you permission to look at the cosmos not with reverence for its mystery, but with respect for its memory.

Let this be the beginning—not of another theory, but of a better way to listen.

—**Sandeep Chavan**
Independent Researcher,
Developer of Universal Energy Dynamics

How to Read This Book

This book is not a traditional scientific manual, nor is it a casual philosophical musing. *Dark ≠ Hidden* is a paradigm shift—meant to be digested, reflected upon, and revisited.

Here's how to approach it:

- **Start with openness.** If you're bound to the belief that light, force, and time explain everything, pause. This book invites you to reconsider those foundations—not through speculation, but through re-alignment with consequence.
- **Don't rush.** The chapters unfold like ripples—each one arising from the last. Let the ideas settle. Think of this not as linear reading but as a field experience.
- **Pause at the diagrams and glossary.** The ripple logic vocabulary is new. Use the glossary frequently. You are not memorizing terms—you're tuning your intuition.
- **Read alone, discuss aloud.** Many sections are structured to challenge the mainstream. Read quietly, then debate with anyone willing. Science grows through friction.
- **Revisit Part IV and the Appendices.** These are not footnotes—they are the experiments, validations, and consequences that future science must grapple with.
- **If you're a scientist—strip your tools of assumptions.** This book doesn't ask you to abandon reason. It asks you to question what your reason is built on.
- **If you're a thinker—welcome back.** UED isn't a new theory. It's a memory you've had all along, waiting to be recalled.

Prologue

Echoes Can't See the Silence

Why modern physics is blind to the true nature of "dark" phenomena—using tools made of the very effects it's trying to decode.

There's a moment in every scientific revolution where the tools of observation themselves become the cage. Where what we call "knowledge" becomes just a feedback loop of previous assumptions. Where light, time, motion, and gravity—once our instruments of truth—become the very reasons we can no longer see what lies beneath.

We are living in that moment now.

Physicists have mapped galaxies with light, measured spacetime with clocks, and interpreted mass through motion. And yet, when confronted with anomalies—like galaxies rotating too fast, light bending where no mass is seen, or distant clusters drifting apart in ways no known force explains—we respond not by questioning the lens, but by inventing the "dark."

Dark matter. Dark energy. Elegant names for desperate placeholders. They aren't discoveries. They're denials.

When galaxies misbehave, we imagine invisible mass. When redshift doesn't align, we summon mysterious repulsion. When gravity refuses to curve space as expected, we bend our equations instead.

And all of this is done using tools derived from the manifestations of a field we haven't even begun to understand.

The Great Epistemological Trap

Let's name the trap clearly:

We are trying to decode Structured Vacuum Energy (SVE) using effects that are *caused* by SVE—light, motion, time, and force.

Imagine trying to study silence using only echoes. Or trying to measure stillness with a seismograph.

Mainstream physics has built its framework on ripple effects—on disturbances in the field—but when it seeks to understand the still field itself, it remains blind. Because it is asking a question in the wrong language.

A Misaligned Inquiry

Dark matter is not "invisible"—it's simply not aligned with ripple-based tools. Dark energy is not "mysterious"—it's just the relaxation of the ripple field, not a force.

Yet modern physics continues to press harder with bigger colliders, deeper space telescopes, more sensitive instruments. But what are they all built upon? Light-based detection. Motion-based timing. Gravitation-based inference. All of them are echoes—none of them can see the silence.

Because the field they're trying to probe—Structured Vacuum Energy—does not ripple in the way these tools expect. It doesn't couple with light. It doesn't emit, reflect, or respond. It simply exists.

And if you're looking for motion in a place of stillness, you'll always conclude something's "missing."

From Illusion to Insight

This book begins with a simple premise:

There is no dark matter. There is no dark energy. There is only Structured Vacuum Energy behaving as it should—and a scientific model that has failed to interpret it.

You will not find exotic particles or speculative repulsive forces here. You will find consequence, gradient, entropy, and alignment—the true mechanics of reality as seen through the lens of Universal Energy Dynamics (UED).

This is not a tweak to existing theories. This is a reset. A shift from believing our tools are universal, to realizing they are contextual consequences.

It's time we accepted the obvious:

You cannot study the origin with its byproducts. You cannot understand silence with sound.

Welcome to *Dark ≠ Hidden*. Let's return to the source—and see clearly, for the first time.

PART I — THE DARKNESS MYTH

Chapter 1: The Invention of Darkness

How "dark" matter and energy became placeholders for failure in fundamental understanding

They didn't discover darkness—they named it when their math stopped working.

When galaxies spun faster than expected, when clusters bent light with more force than their visible mass could explain, and when distant stars whispered a redshift no theory predicted, physicists stood at the edge of a terrifying truth:

Their understanding of the universe was incomplete.

But instead of confronting that, they did something far more convenient. They called it *dark.*

Dark matter. Dark energy. Elegant titles for what amounts to a scientific shrug.

For centuries, science claimed progress through illumination. But here, it took a different turn. Faced with anomalies, it didn't search deeper—it cast shadows. What couldn't be explained was boxed, labeled, and archived under darkness, not because it was hidden, but because existing frameworks refused to bend.

"If the stars won't obey our gravity, let's invent mass they can obey."
"If the universe stretches too fast, let's imagine a force pushing it."

These weren't solutions. They were silences. And the tragedy is not that we don't know what dark matter or dark energy is— The tragedy is that we never asked if we were looking the wrong way.

In this chapter, we'll revisit the exact moments where modern cosmology took a detour— Not because reality failed to cooperate, but because physics mistook its own echo for a signal.

This is not a story of missing things. This is a story of misunderstood consequences.

Welcome to the invention of darkness. Not a mystery of the cosmos—but a mirror held to our own blind confidence.

The Rise of the Placeholder

Physics, at its best, is a language of precision. But at its weakest, it becomes the art of naming what it cannot explain. And in no area has this been more apparent than in the birth of "dark matter" and "dark energy"—terms that sound profound, but are functionally nothing more than placeholders.

They weren't discoveries. They were corrections.

The Moment Gravity Broke

In the 1930s, astronomer Fritz Zwicky was studying the Coma galaxy cluster when he noticed something alarming: the galaxies were moving far too fast. Based on Newtonian and Einsteinian models, these galaxies should have flown apart—but they didn't. Zwicky concluded that there must be some invisible mass holding them together.

Thus, "dark matter" was born—not through observation, but through inference.

There was no direct detection. No interaction. Just a gravitational mismatch—and a label to account for it.

Decades later, Vera Rubin observed a similar anomaly in the rotational speeds of individual galaxies. The outer stars in spiral galaxies were rotating too quickly. Again, rather than questioning the gravitational assumptions, scientists doubled down: more dark matter must be there. Not because it was seen, but because the math demanded it.

The Crisis Deepens: Enter Dark Energy

In 1998, two independent teams studying distant supernovae came to an even more shocking conclusion: the universe's expansion was accelerating. According to classical cosmology, it should have been slowing down.

To account for this, physicists revived an old ghost from Einstein's original equations—the cosmological constant—and rebranded it as "dark energy."

This wasn't a measurement of energy. It was a forced patch—a pressure plug inserted into the framework to hold the universe together.

Just like dark matter, dark energy was a placeholder—a name invented not from clarity, but from necessity.

"When the universe refused to collapse, we invented something to push it farther apart."

From Epicycles to Darkness

This isn't the first time science has played this game.

In the age of Ptolemaic astronomy, geocentric models reigned supreme. To explain why planets didn't move in simple circles around Earth, scientists invented epicycles—little imaginary loops that planets were said to trace. The more data poured in, the more epicycles were added. The model grew *more precise, but less real.*

It wasn't until Copernicus and Kepler that the deeper truth emerged:

The framework itself was wrong. The problem wasn't the planets—it was the perspective.

Dark matter and dark energy are today's epicycles. Not because they're inherently false—but because they are born of desperation to preserve a crumbling worldview.

Physics by Patchwork

Real discoveries expand understanding. Placeholders freeze it.

When science starts naming gaps instead of bridging them, it loses sight of its purpose. Dark matter and dark energy aren't mysteries to be solved. They are signals—not of something missing in the universe, but of something missing in our understanding.

"When the numbers stopped working, we added shadows to preserve the light." But shadows are not entities. They are consequences of blocked perception.

This section reveals the uncomfortable truth: Darkness, as used in modern cosmology, is not a physical reality. It is a theoretical crutch—a mathematical bandage wrapped around a broken foundation.

In the next section, we will examine that foundation more closely. Because to understand why we invented darkness, we must look at the instruments of our blindness.

The Empirical Panic

Science is built on observation—but when observations begin to challenge the foundation, the response is often not reform, but retreat.

Modern cosmology faced three such moments—three deep jolts to its confidence—where the universe refused to behave as predicted. And instead of reevaluating the core assumptions, physicists clung tighter to their frameworks, patching the models instead of rethinking them.

This section explores those turning points—where reality knocked, and science bolted the door shut.

The First Crack: Zwicky and the Galaxy Clusters (1933)

In the early 1930s, Swiss astronomer Fritz Zwicky was studying the Coma cluster, a massive group of galaxies. Using the virial theorem (a balance of kinetic and potential energy), he calculated the expected gravitational pull required to hold the galaxies together.

But the math didn't add up.

The visible matter—the stars, gas, and dust—wasn't enough. According to Newtonian physics and even Einstein's general relativity, these galaxies should have drifted apart at the speeds they were moving.

So what did Zwicky conclude? That there must be some invisible mass, an unseen substance holding the cluster together. He called it "dunkle Materie"—dark matter.

It wasn't observed. It wasn't measured. It was inferred from a mismatch between observation and assumption. And with that, the first placeholder was born.

The Confirmation Bias Deepens: Galactic Rotation Curves (1970s)

Decades later, astronomer Vera Rubin and her team were mapping the rotational speeds of stars in spiral galaxies. According to known gravitational laws, stars farther from the galactic center should rotate more slowly—just as outer planets in the solar system do.

But they didn't.

In Rubin's data, stars in the outer arms of galaxies were rotating just as fast as inner ones. Once again, the observable matter couldn't account for this.

And again, the same reasoning emerged:

"There must be more mass. We just can't see it."

No one asked whether the laws themselves—the equations used to predict gravitational pull—might be inadequate at galactic or intergalactic scales.

Instead, the concept of dark matter gained momentum—not as a discovery, but as a plug to keep the gravity model intact.

"Instead of questioning the curvature, we invented the weight."

The Final Blow: Accelerating Universe (1998)

In the late 1990s, two research teams—one led by Saul Perlmutter, the other by Brian Schmidt and Adam Riess—were observing distant Type Ia supernovae to measure the expansion of the universe.

They expected to confirm that cosmic expansion was slowing down due to gravity.

But they didn't.

Instead, the data showed something truly unexpected: The expansion was speeding up.

This shook the very foundation of cosmological physics. Under Einstein's model, there was no mechanism to explain accelerating expansion without invoking something deeply unnatural—some form of anti-gravity.

And so, just like Zwicky and Rubin before them, the physicists improvised:

"There must be a repulsive energy permeating space." "Let's call it... dark energy."

This wasn't a discovery—it was a forced reconciliation. The universe's behavior was screaming for a new paradigm, and science responded with conceptual duct tape.

Panic over Paradigm

In all three cases, reality offered an invitation:

"Come closer. Reconsider what you think you know."

But physics responded with fear. The response wasn't: "Let's question gravity." It was: "Let's invent what the equations demand."

Instead of recognizing that their model of gravity might be incomplete—or their understanding of light, time, and energy might be ripple-derived—physicists invented substances and forces to account for every misfit in the data.

Not a single one of these "dark" entities was ever directly observed. Not one has been experimentally isolated or validated. They exist only to protect the framework from collapse.

The Role of Mathematical Insistence

Einstein's general relativity taught us to think of gravity as the curvature of spacetime caused by mass-energy. But once this elegant metaphor took hold, it became a dogma.

Now, if the curve is steeper than expected, science doesn't question the nature of the field—it assumes more mass must be present.

"If gravity curves spacetime, we must find what's pulling."

And if the curvature begins to *relax* or diverge faster than expected? We assume there's a cosmic push—an energy force expanding space itself.

But what if curvature isn't the right language at all?

What if space is not a fabric, and gravity is not a force?

What if the entire phenomenon is ripple behavior—an emergent consequence of energy gradient and alignment across a still field?

This is where **Universal Energy Dynamics** steps in—not to *add another patch*, but to *remove the illusions altogether.*

In the next section, we'll expose the instruments of blindness—the very tools by which modern science has misread the cosmos. Light, gravity, and time: the ripples that deceive when mistaken for truth.

The Tools That Lied

How our instruments became our blindfolds in the search for truth.

Modern physics prides itself on being empirical—on seeing, measuring, and validating theories against observable data. But what happens when the very tools of observation are biased toward a particular kind of reality?

What if they only detect *motion*, *light*, and *force*—and are blind to what lies beneath?

This is not a philosophical question. It is the exact reason Structured Vacuum Energy (SVE) has gone unnoticed, misunderstood, and mislabeled as "dark matter" and "dark energy."

The Ripple Bias

Every tool we use in physics is ripple-biased.

- Light-based observation depends on electromagnetic wave

interactions.

- Gravitational inference assumes mass creates spacetime curvature through ripple-like stress on geometry.
- Motion tracking uses displacement over time—a ripple-derived phenomenon dependent on perceived change.

But all of these rely on a common feature:

They detect only when ripple alignment takes place.

No alignment, no signal. No ripple, no interaction. No interaction, no detection. And from this absence of signal, we infer absence of reality.

This is the root fallacy of modern cosmology.

Looking With Echoes

Imagine trying to observe stillness with a microphone. Or trying to detect silence by blasting sound and waiting for reverberation.

You won't hear the quiet. You'll hear your own reflection.

This is what science has done to the cosmos.

It looks at the universe with tools designed to capture movement, energy shifts, photon transactions—and when nothing responds, it labels that zone "dark." Not because it is untrue, but because it refuses to echo back.

"You were using motion to find the unmoving."

The Ripple Consequence Chain

Let's break this down clearly:

1. All matter, as understood by physics, is condensed ripple activity in the vacuum field.
2. Light is a ripple-based transfer—an alignment event across structured vacuum layers.
3. Gravity is perceived only where energy gradients create ripple response.
4. Time and motion are subjective experiences based on ripple delay and phase propagation.

Thus, everything we "observe" is not primary reality—it is ripple behavior. We've built a science on echoes, not on the silent source.

The Real Reason SVE "Doesn't Respond"

Structured Vacuum Energy—especially in its stable, non-condensed form—does not ripple in a way detectable by these instruments.

- It doesn't emit or reflect light—because no electromagnetic alignment occurs.
- It doesn't resist motion—because it holds no condensed tension.
- It doesn't cause ripple acceleration—because it is already in equilibrium.

Hence, it appears invisible. Untouchable. Missing.

But it isn't missing. It simply doesn't ripple back.

The False Conclusion

When our detectors return silence, we conclude something is absent.

But in UED, silence means something very different:

The field is aligned. The ripple is at rest. No interaction is required.

Dark matter appears "invisible" not because it hides—but because it is ripple-silent. Dark energy seems undetectable not because it's ethereal—but because it is ripple-diffusing, not reflecting.

Ripple-Awareness: The UED Insight

The tools didn't lie intentionally. They just weren't built for the job.

You cannot study SVE using light-based instruments, because light itself is a ripple product of SVE. You cannot measure fundamental stillness with tools made to detect motion. This is why science keeps chasing "darkness"—and never finds it.

"Absence of ripple response is not absence of presence—it is presence in perfect equilibrium."

As we move into the next section, we confront the heart of the error—not just in measurement, but in worldview. The mistake wasn't in the tools alone. It was in the belief that those tools were universal.

Let's now examine how this blindness became institutionalized—how curvature became gospel, and anything that broke it was patched, not questioned.

The Philosophical Collapse

Why "darkness" in physics reveals more than missing mass—it exposes a failure in how we think about reality itself.

The search for dark matter and dark energy is not just a scientific pursuit. It is a philosophical emergency in slow motion.

For decades, physicists have operated with quiet confidence that the universe can be understood through interaction: Light strikes, matter bends, time ticks, forces pull.

But dark matter doesn't strike. Dark energy doesn't push. And yet, they still shape the cosmos.

This should have prompted a rethinking of what counts as real. Instead, science doubled down.

Mistaking the Effect for the Cause

At the heart of the philosophical collapse is a simple but devastating mistake:

Modern physics mistook the ripple for the source.

Everything it studies—light, time, gravity, motion—is a consequence of deeper field dynamics. These are not the building blocks of reality. They are the outcomes.

But physics, trapped in its own echo chamber, took these outcomes as the fundamentals:

- Gravity was not seen as ripple alignment—it was declared a force.

- Light was not treated as a ripple transaction—it was idealized as a universal messenger.
- Time was not understood as ripple delay—it was elevated to a fourth dimension.
- And when none of these tools could explain the galactic anomalies, science invented *dark* versions of mass and energy—still clinging to the framework that caused the confusion.

It's like watching a puppet move, and trying to explain its behavior without acknowledging the puppeteer.

Absence of Interaction ≠ Absence of Existence

The second, subtler failure is the assumption that if something doesn't interact, it must not exist.

This is the unspoken rule behind every telescope, detector, and theory in mainstream science: "If we can't observe it, it's either not real or not relevant."

But Structured Vacuum Energy (SVE)—the still field that underlies all ripple-based phenomena—doesn't want to interact. It is not designed to reflect, push, or pull. It simply holds the architecture of reality in stillness. That's why it doesn't appear in any equation that expects a response.

SVE doesn't bend spacetime. It doesn't produce light. It doesn't generate force. It just *is*.

And because it *is* without *doing*, physics has ignored it—despite it being the very foundation of every ripple, every motion, every observable event in the universe.

The still field has been dismissed, not because it's false, but because it's quiet.

UED: A Framework Beyond the Chase

Universal Energy Dynamics (UED) doesn't chase what's missing. It doesn't hunt for unseen mass or repulsive forces.

Instead, it asks:

- Why are we seeing anomalies in ripple behavior?
- What field conditions would naturally produce these effects without inventing new entities?
- What does it mean for something to be "real" if it doesn't ripple back?

In UED, dark matter is not matter. Dark energy is not energy. Both are simply the logical outcomes of ripple behavior across a still, structured vacuum field.

Rather than naming shadows, UED traces the origin of the shadow. Rather than adding mass to equations, it removes the assumption of mass as fundamental. Rather than measuring acceleration, it tracks entropy-driven ripple dissipation.

A Shift in Ontology

The implication is profound:

Maybe the universe isn't hiding anything. Maybe our framework just isn't aligned with what's actually there.

This is not a call to abandon science—it's a call to refine our understanding of what science is truly observing.

Maybe the problem isn't the cosmos. Maybe the problem is how we define causality.

Closing Teaser

So here's the question we end with:

What if there's nothing missing? What if your equation is simply misaligned with the field that never left?

Because when we finally stop naming ghosts... When we stop trying to force the universe to echo back...We might just hear the stillness we've been standing in all along.

With this realization, we now move forward—not into deeper darkness, but toward the field itself.

Chapter 2: Physics in a Mirror

They built equations on what ripples show—not on what the field is.
The error of studying the vacuum with its own manifestations.

Imagine a scientist trying to understand the ocean by studying only the waves—never diving beneath the surface, never questioning what lies below. Every conclusion is drawn from what moves, what crashes, what reflects the sun.

This is exactly how modern physics has treated the universe.

From Newton's falling apples to Einstein's curved spacetime, our greatest theories have all been constructed by observing ripples—by decoding motion, light, force, and time. But these are not the source of reality. They are the manifestations—consequences of a deeper field we never bothered to study directly.

Physics has been looking into a mirror—watching reflections of activity, building equations from echoes. But it never turned around to examine the source that produced those echoes.

We studied the shine, but never the lamp. We measured the ripples, but never the stillness they disturbed.

And when those ripples didn't behave as expected—when galaxies spun too fast, or light bent too sharply—what did we do?

We blamed the mirror. We blamed the universe for not matching the reflection we predicted.

This chapter confronts that error.

We will explore how science became obsessed with ripple behavior while remaining blind to the structured vacuum field that governs all ripple initiation. We will see how our equations reflect reality only partially—because they are drawn not from the origin, but from after the fact.

And we will begin to lay the groundwork for an entirely new way of seeing: A physics that doesn't just describe what ripples do—But understands why they begin at all.

Welcome to the other side of the mirror.

The Mirror Mistake

They measured the echoes and called it the voice.

The triumph of modern physics is built on patterns. Patterns of light, patterns of motion, patterns of gravitational pull, charge, mass, time. And for centuries, we have celebrated this ability to describe and predict what these patterns will do. But what if that entire pursuit was based on a fundamental oversight?

What if the patterns we measure are only consequences—and not the origin?

Ripple Consequences, Not Source Mechanics

At its foundation, physics has always been a ripple-centric science. Whether it's Newton's force diagrams, Maxwell's wave equations, or Einstein's spacetime geometry, every formulation is derived from what happens once energy begins to move. Once a system ripples—visibly, measurably—then and only then does physics begin to pay attention.

But what causes that ripple to emerge in the first place? What decides *when* and *where* energy will manifest into form, motion, or radiation?

That question has remained outside the equation.

Science doesn't ask why the wave began—only how fast it moves, how far it travels, how it interacts. But this makes modern physics a mirror: it only reflects what has already occurred. It doesn't see the still field before the disturbance.

The Analogy: Studying Fire by Measuring Only Smoke

Imagine a fire. It produces heat, light, and smoke. A ripple of consequence spreads outward. Now imagine a team of scientists trying to understand fire—but they only measure the smoke.

They catalogue its color, thickness, direction. They build models to predict its rise and fall, its diffusion pattern. But they never examine the flame. They never trace the ignition.

This is exactly what we have done with gravity, light, motion, and time. We observe the smoke of the ripple—but we've never truly understood the energetic ignition of the ripple field.

The equations that built physics are not wrong. But they are echoes, not the voice.

From Description to Deception

Here's the real danger: when ripple equations begin to fail—when they no longer predict what galaxies or particles actually do—science doesn't question the ripple-centric worldview. It adds complexity to the mirror:

- Extra dimensions.
- Dark matter.
- Dark energy.
- Quantum probability clouds.

Instead of turning around to examine the source field, physics polishes the reflection, hoping a sharper image will make the truth clearer.

But a mirror, no matter how clear, can never show what is behind it.

Ripple Bias as Structural Error

This is not just a technical flaw. It's a philosophical mistake embedded into the foundation of scientific reasoning:

We assumed that what we can measure is all that exists. And we assumed that measurement is only valid when interaction occurs.

In Universal Energy Dynamics (UED), this assumption is rejected.

The true origin of light, mass, and motion lies not in the equations they produce—but in the conditions of the vacuum field that allowed ripple alignment to occur in the first place. Without this foundation, all equations are after-the-fact reconstructions of partial truths.

Physics has never described the universe itself. It has only described the manifested response of the universe under ripple stress.

Closing Reflection

We must now step away from the mirror. Not because it lied to us— But because it was never the whole room.

Let us begin asking different questions: What exists before the wave? What field holds the ripple before it forms? And why does it sometimes remain still, refusing to echo back?

Because until we ask these, we will continue measuring smoke—While the fire burns somewhere beyond the frame.

Foundations Made of Reflections

We built the palace of physics on the behavior of echoes, not the presence of the voice.

From the earliest classical formulations to the most abstract modern theories, physics has constructed its worldview on the aftermath of events—on motion already begun, forces already expressed, and ripples already in play.

The great irony? The most celebrated equations in science are not descriptions of what the universe *is*—but of what it *does after something disturbs it.*

These are not blueprints of being. They are postmortems of movement.

Newton's Laws: Describing After-the-Fact Interactions

Take Newton's laws of motion and universal gravitation. These were revolutionary in their ability to describe what happens when a force acts on a body:

- Objects at rest stay at rest (unless something disturbs them).
- Force equals mass times acceleration (after interaction begins).
- Every action has an equal and opposite reaction (a reaction to something already applied).

Nowhere in Newton's mechanics is there an attempt to understand:

- What initiates the force?
- Why mass resists motion?
- What mass fundamentally *is* before the force interacts?

Newton's framework begins only when motion begins. It's a science of consequences, not causes.

Maxwell's Equations: Ripple Codified

Maxwell's equations are often praised as the crown jewel of classical physics. They describe the behavior of electric and magnetic fields, and how they give rise to electromagnetic waves—light itself.

But again, this is ripple analysis.

Maxwell doesn't tell us where the field comes from, why it exists, or what condition of space permits a disturbance to move at the speed of light. His system beautifully captures how the wave propagates—*once it's already been launched.*

What Maxwell gave us was not the origin of light, but the grammar of its ripple.

Einstein's Relativity: The Geometry of Disturbed Fields

Einstein's general relativity was a leap beyond force—it introduced the idea that mass-energy curves spacetime, and that objects follow the geometry of that curvature.

Elegant? Yes. But again, ripple-centric.

Spacetime curvature is a field response to the presence of condensed energy. There is no account of:

- Why the vacuum allows curvature in the first place
- What the vacuum *is* without mass-energy
- How motion or time emerges prior to disturbance

Einstein modeled the geometry of disturbance, not the nature of stillness.

And when that curvature didn't match observations (e.g., galaxy rotation, cosmic acceleration), what did science do?

It introduced dark matter and dark energy—not by rethinking the field, but by doubling down on the curvature. A deeper consequence of the same mistake.

A Model of Consequences, Not Conditions

Let's be clear:

- Newton told us what happens after a force is applied.
- Maxwell told us how fields ripple once energy is introduced.
- Einstein told us how spacetime curves after mass exists.

But none of them told us:

What exists before any of this happens.

They never defined the conditions of initiation—the stillness that must exist and be disturbed for any of their equations to activate.

Effect Chasing Effect

This is the recursive loop modern physics is trapped in:

1. Observe a ripple (light, motion, acceleration).
2. Describe it with a formula.
3. Wait for another ripple.
4. Patch your formula.
5. Repeat.

But at no point does it ask:

- What allowed the ripple to occur in the first place?
- Why do some fields ripple and others remain silent?
- What governs whether alignment happens at all?

This is not a physics of origin. It is a physics of reaction.

The UED Critique

Universal Energy Dynamics identifies this as the core flaw in the mainstream model:

Physics has become a system of ripple-recognition, not field-understanding.

It's like hearing thunder and trying to explain weather patterns without ever noticing the cloud.

The ripple is only the messenger. The field is the author.

Until we shift our equations from consequence to condition, from response to readiness, we will keep running in circles—chasing effects, labeling absences, inventing darkness where only stillness remains.

In the next section, we turn our attention to that neglected presence—To the Structured Vacuum Energy that holds all ripple potential, yet remains unmeasured by any tool that expects an echo.

The Vacuum Misunderstood

"They saw a pause between events and called it emptiness."

The term "vacuum" has always occupied a strange space in physics—defined not by what it is, but by what it *lacks*. It is, by default, a placeholder for absence. An empty backdrop. A null zone.

In classical mechanics, the vacuum is quite literally "nothing." In quantum physics, it is a boiling soup of fleeting particles—chaotic, unstable, and unpredictable.

Both views are wrong. Both are projections of ripple-biased thinking onto a field they fundamentally do not understand.

The Classical Mistake: Vacuum as Nothing

In Newtonian and pre-modern physics, the vacuum is simply the absence of matter—a passive container in which forces act and objects move. It has no role in reality except as background space.

This view was deeply entrenched in Western science: If nothing is there, then nothing *is*.

So when waves disappeared, particles ceased, or forces dropped to zero, physicists concluded:

"We've entered a vacuum."

But this wasn't observation. It was assumption. They mistook the pause between ripples for the non-existence of the field itself.

They believed the stage was empty when, in truth, it was simply quiet.

The Quantum Overcorrection: Vacuum as Chaos

Quantum field theory tried to fix this—but swung too far.

In quantum physics, the vacuum is not empty. Instead, it is restless—buzzing with virtual particles, zero-point energy, and probability storms. Here, "nothing" becomes a statistical frenzy—particle-antiparticle pairs appearing and vanishing in Planck-scale blinks.

But again, this is a ripple-centric interpretation.

Quantum physics saw fluctuations in measurement and declared that the vacuum itself was unstable. But what if those fluctuations are not in the vacuum—but simply disturbances of it?

What if the vacuum isn't chaotic at all—but simply misread through ripple-sensitive instruments?

In both classical and quantum frameworks, the vacuum is misunderstood because the measuring stick itself is ripple-biased.

Enter SVE: The Structured Vacuum Energy

Universal Energy Dynamics proposes a radically different view:

The vacuum is not empty, and it is not unstable. It is Structured Vacuum Energy (SVE)—a still, coherent, energetically balanced foundation. It does not emit. It does not interact. But it holds all ripple potential.

SVE is not "space." It is not a container. It is the energetic medium from which all phenomena emerge—and into which they dissolve.

It is not passive. It is pregnant with ripple possibility—but only activates when alignment thresholds are met.

In UED terms:

- The vacuum does not ripple unless disturbed.
- And when it does ripple, it behaves with topological consistency and gradient-seeking intelligence.

Ripple as Reaction, Not Action

This is the key philosophical shift:

Ripple events are not actions happening *in* the vacuum—they are the vacuum's *response* to imbalance.

Just like water doesn't move unless agitated, SVE remains still until:

- Energy gradient is introduced,
- A boundary condition is applied,

- Or misalignment forces the field into restoration mode.

In other words, the vacuum is not where things happen—it is what happens when things are about to happen.

Ripple is a reaction of the field—not a substance within it.

From Emptiness to Readiness

So why has this been so difficult for mainstream physics to accept?

Because it mistook:

- The absence of particles for the absence of energy,
- The lack of ripple for the lack of presence,
- The invisibility of SVE for non-existence.

"They saw a pause between events and called it emptiness."

But SVE is not nothing. It is the first condition of everything. It is what makes light, gravity, and time even possible.

And this is why "dark matter" and "dark energy" appear invisible—because our ripple-biased tools interpret this still, non-interacting field as *nonexistent* when in truth, it is simply non-rippling.

In the next section, we pull this perspective into focus with a powerful conclusion: Modern physics has built its world model on ripple effects—but has never truly studied the field that causes them.

It's time to flip the paradigm.

From Symptom to Source: The UED Shift

"Physics looked in a mirror and saw patterns. UED turns around to see what's making the mirror."

The story of physics so far has been a story of ripple diagnosis. It observes a symptom—motion, force, curvature, light—and attempts to reverse-engineer its cause. But it never questions the conditions that allowed the symptom to emerge in the first place.

This is the core pivot offered by Universal Energy Dynamics (UED): It shifts the question from *What happened?* to *What allowed anything to happen at all?*

From Post-Event Math to Pre-Event Insight

In conventional physics, equations are applied after an observable event:

- A force is applied, so we calculate motion.
- A wave is emitted, so we measure frequency.
- A mass curves spacetime, so we compute the geometry.

But UED takes a different route. It asks:

"What *conditions* had to be in place for this ripple to even begin?"

Rather than solve equations after manifestation, UED traces the ripple's origin back to:

- The stability of the vacuum.
- The misalignment in field equilibrium.
- The gradient that initiated the ripple event.

Ripple is not a default. It is a response to imbalance—and UED focuses on what destabilizes stillness, not what happens once it's already in motion.

The Core Drivers: Gradient, ATT, and Entropy Field Logic

UED introduces a new vocabulary and causal structure beneath all physics:

Ripple Gradient

The difference in energy density across SVE that initiates motion—not through force, but through compensatory realignment.

This replaces the need for mysterious attraction or force carriers. Objects don't pull—they adjust within their local ripple environment based on gradient dynamics.

ATT – Alignment Threshold Time

This is the minimum temporal condition required for two systems to synchronize their ripple behavior.

If no ATT is achieved, no interaction occurs. This is why SVE zones (like those we call "dark matter") do not respond to light—they do not meet ATT conditions for electromagnetic coupling.

ATT is the missing causal layer beneath what we perceive as reflection, emission, and measurement.

Entropy Field Dynamics

Ripple is governed by tension and release. Every ripple event is a local entropy redistribution—trying to restore equilibrium to the SVE field.

Instead of forces and particles, UED reads the universe as a dance of entropy gradients—with ripple behavior as the music of imbalance seeking harmony.

Causality Before Kinetics

This approach changes everything.

It means:

- Motion isn't the result of force—it's the resolution of misalignment.
- Light isn't a default emission—it's a ripple agreement between systems.
- Time isn't absolute—it's a perception of ripple sequencing

when alignment occurs.

Mainstream physics operates on event-response logic. UED operates on field-readiness logic.

We're no longer treating the fire after the smoke appears. We're understanding what conditions make the spark possible.

Stepping Behind the Mirror

Modern physics was content analyzing patterns—ripples on the surface, equations tracking motion, light, force, and energy like shadows across a wall.

"Physics looked in a mirror and saw patterns."

But UED breaks the frame.

It turns around to study what's making the mirror reflect at all—The still field, the structured vacuum, the thresholds of ripple, and the causality embedded in silence.

This is not a refinement of old theories. It is a re-rooting of physics into its actual origin—the field before the wave.

Chapter 3: The Dark Matter Mirage

Why galaxies seem to rotate too fast—and why the "invisible mass" solution is an illusion of ripple anchoring.

The stars wouldn't slow down.

In the quiet arms of spiral galaxies, far from the dense gravitational centers, stars were spinning at astonishing speeds—too fast to be held by the visible matter around them. According to Newton and Einstein, they should have been flung into the void long ago.

But they stayed. And with that, the illusion began.

To explain this, physicists conjured an elegant specter: Dark Matter—a form of invisible mass, undetectable by light, touch, or trace, yet somehow comprising over 80% of the universe's total matter content.

It was a fix, not a find.

Because the math didn't work, because the curves didn't fall, because the equations refused to match observation, we assumed the cosmos must be full of something we couldn't see—and in doing so, we declared the unseen to be real, simply because our theory demanded it.

But what if the stars weren't misbehaving? What if gravity wasn't failing? What if the field was working perfectly—but we were asking the wrong question?

This chapter challenges the most accepted explanation in astrophysics. Not by denying the data—but by redefining what the data truly shows.

We will discover that the need for "invisible mass" stems not from missing matter, but from misunderstood ripple mechanics. That what holds galaxies together is not mass hidden in shadow, but Structured Vacuum Energy in perfect ripple equilibrium—anchoring stars not with pull, but with alignment.

Dark matter is not a thing. It is a misunderstanding born of ripple blindness.

Let's dismantle the mirage. The galaxies were never lying. Our framework was.

The Galactic Anomaly That Shook Gravity

"The stars refused to slow down, so we invented something to hold them back."

In the early 20th century, astronomy reached further into the cosmos than ever before. Telescopes began resolving galaxies not just as hazy smudges but as vast spirals of stars rotating in organized patterns. And then something strange was noticed—something so persistent, so universal, that it eventually challenged the very core of gravitational theory.

The stars on the outer edges of galaxies were spinning too fast.

The Flat Rotation Curve Mystery

According to Newtonian gravity—and later supported by Einstein's general relativity—the orbital velocity of stars should decrease the farther they are from the galactic center. This follows the same logic as our solar system, where Mercury moves fastest and Neptune crawls at the edges.

But when astronomers began charting the actual velocities of stars in spiral galaxies, they found a different story:

- Inner stars behaved as expected.
- But beyond a certain radius, the rotation curve flattened—stars were moving at roughly the same speed as those closer in.

This was not a one-off anomaly. It was everywhere.

From the Andromeda galaxy to distant spirals billions of light-years away, stars in the outskirts were spinning with reckless confidence, unshaken by the apparent lack of gravitational support.

The implication? According to existing physics, these galaxies should be flying apart. But they weren't.

Vera Rubin and the Invention of "Missing Mass"

In the 1970s, astronomer Vera Rubin and her colleague Kent Ford conducted a comprehensive survey of galaxy rotation curves. Their work confirmed and expanded the anomaly across many different galactic systems.

Faced with the consistency of this contradiction, Rubin reached the only conclusion allowed within the existing gravitational framework:

"There must be more mass than we can see."

This "missing mass" had to be distributed in vast, invisible halos surrounding every galaxy. It didn't emit light. It didn't absorb radiation. It didn't collide. But it had to exist—because otherwise, the math wouldn't work.

Thus, dark matter became official.

Not as a measured entity. Not as an observed reality. But as a necessary invention to save gravitational theory from collapse.

The Undetectable Problem

Since then, decades of research have gone into trying to directly detect this hypothetical substance. Billions have been spent on dark matter detectors buried deep underground—shielded from radiation, calibrated to spot a single interaction from a "Weakly Interacting Massive Particle" (WIMP), or a "supersymmetric relic."

The result?

Nothing. Not one particle. Not one trace. Not one verified interaction.

And yet, dark matter remains *in the model*—propped up not by evidence, but by necessity.

Because without it, the equations break. Because if the mass isn't invisible, then gravity itself must be wrong.

The Real Issue: The Theory Was Too Fragile to Fail

The idea of "missing mass" was never born from observation—it was deduced from failure. A model couldn't explain star motion, so the stars were blamed. The universe was asked to carry invisible baggage so that physics could avoid self-reflection.

The stars refused to slow down, so we invented something to hold them back.

This is not science at its best. This is science in survival mode.

The galactic anomaly should have been a signal—an invitation to explore a deeper field reality. Instead, it became the birthplace of a myth—a ghost story we turned into doctrine.

In the next section, we'll see how this entire line of thinking stems from a fundamental misread—not of data, but of ripple mechanics. The field didn't fail. Gravity didn't collapse. The stars weren't misbehaving.

Our interpretation was.

Why Gravity Failed—The Ripple Misread

"They saw speed, and assumed pull. They missed the field that sustains flow."

The mystery of dark matter arises not because the universe is hiding something, but because physics is looking with the wrong eyes.

Gravity, as it is currently understood, is a mathematical structure built to describe how point masses attract each other through space. Whether in Newton's formulation as a universal force, or in Einstein's reimagining as spacetime curvature, the core assumption remains unchanged:

Objects attract each other due to their mass, and that attraction diminishes with distance.

It's a tidy, classical framework that works well—on small scales, in isolated systems, or near dense masses.

But galaxies are not small. They are not isolated. And they are not mechanical systems of point masses.

Gravity as a Localized Force

In both Newtonian and relativistic models, the gravitational effect of an object is strongest near its center and falls off with distance. This is why in our solar system, planets farther from the Sun orbit more slowly: the force weakens with distance squared.

So naturally, physicists expected stars on the outskirts of galaxies to behave the same way.

But they didn't.

Why?

Because galaxies are not collections of marbles spinning in space. They are immersed in a ripple field, and their behavior emerges from ripple dynamics, not mechanical pull.

Galaxies as Ripple Environments

In Universal Energy Dynamics (UED), mass is not a substance exerting force. It is a zone of condensed ripple within Structured Vacuum Energy (SVE).

The space between stars is not empty—it is a ripple-ready field, still coherent, still structured, and still engaged with the mass zones around it. Galaxies are not gravitational systems held together by "pull," but energetically connected environments maintained by ripple alignment across SVE.

The stars are not orbiting because they are being pulled. They are orbiting because they are anchored in a shared ripple system that spans the galactic field.

This anchoring creates phase consistency—a dynamic equilibrium maintained not by force, but by ripple continuity. Like dancers moving in rhythm, not because they are tied by ropes, but because they are aligned with the same beat.

Ripple Anchoring vs. Gravitational Containment

Let's draw the contrast clearly:

Model	Mainstream Gravity	UED Ripple Anchoring
Assumes	Point-mass attraction (local force)	Field alignment and ripple continuity across SVE
Mechanism	Force acts through space	Ripple seeks equilibrium through structured field
Between stars	Empty or inert space	Coherent, phase-capable structured vacuum energy
Result	Stars must be pulled by unseen mass	Stars are supported by uninterrupted ripple connections

"They saw speed, and assumed pull. They missed the field that sustains flow."

The Role of SVE Between Stars

The vast space between stars is often assumed to be empty vacuum. But in UED, it is Stable Structured Vacuum Energy—a still, silent medium that holds ripple tension without manifesting mass or light.

This SVE offers ripple support—not resistance or push, but energetic phase continuity. The stars are in motion not against it, but within its coherence.

It's not that outer stars are defying gravity. It's that gravity is the wrong language for what's happening.

UED interprets this not as a gravitational collapse, but as a misread of ripple geometry:

- No extra mass is needed.

- No exotic matter is required.
- The rotation curves are correct—but the explanation has been wrong.

Fluid Analogy: River Flow vs. Rope Pull

Think of a leaf drifting down a river.

A classical physicist might imagine the leaf is being "pulled" by an invisible rope downstream. But in reality, the leaf moves because it's immersed in a flow field—a dynamic medium guiding its motion.

In the same way, stars in a galaxy aren't being yanked by invisible mass—they're flowing within ripple-structured space, held not by pull, but by participation in an anchored field.

A Ripple-Based Interpretation of the Galaxy

Dark matter is not needed because the observed motion is not anomalous under UED. It is the expected result of ripple anchoring across structured vacuum.

This field:

- Does not emit light.
- Does not respond to electromagnetic tests.
- Does not resist movement.

But it aligns ripple behavior across mass zones, enabling rotation stability without requiring additional gravitational pull.

This is the ripple misread—gravity was looking for mass where there was only phase-preserving stillness.

In the next section, we introduce this field more formally—Stable SVE—and show why it's been mistaken for "dark matter" all along.

The mystery was never about what's missing. It was about what was already perfectly present, yet unread by the tools we trusted most.

Stable SVE: The True Anchor

"Dark matter is not hidden mass—it is undisturbed structure."

The entire dark matter narrative hinges on a simple expectation:

If galaxies rotate in ways that defy gravity, there must be more mass than we can see.

But this expectation assumes that the only thing capable of stabilizing motion is mass. That only a gravitational pull from matter—whether visible or invisible—can explain the cohesion we observe.

Universal Energy Dynamics (UED) challenges that assumption at its root. It introduces a new concept that rewrites what holds galaxies together: Stable Structured Vacuum Energy, or Stable SVE.

What Is Stable SVE?

Stable SVE is the silent majority of the cosmos. It is not matter, not radiation, and not "dark" in any mysterious sense. It is still, ripple-ready, structured energy—perfectly aligned with the universe's natural equilibrium.

Think of it as:

- Energetically balanced
- Non-reactive to light
- Phase-consistent with its surroundings

It's not "doing" anything observable—but it's holding everything in place. It is the foundation upon which galaxies float, move, and maintain their structure.

Where physics sees empty space, UED sees ripple support zones—not acting through gravitational pull, but through coherent ripple alignment.

No Mass, No Problem

Here's the critical insight:

You don't need mass to create structure—just field stability.

Mass is a condensed form of ripple energy. But ripple doesn't always condense. Sometimes, it remains in phase-ready equilibrium—coherent, but not interactive. This is Stable SVE.

And yet, because it:

- Doesn't radiate,
- Doesn't reflect,
- Doesn't interfere,

...it gets ignored. Or worse—it gets labeled "missing mass."

But Stable SVE is not missing. It's exactly where it should be—undisturbed, uncompressed, and perfectly maintaining the ripple geometry of the galaxy.

This is the true anchor. Not force. Not gravity. Field-based continuity.

Why It Looks Like "Dark Matter"

Stable SVE has been hiding in plain sight—not because it's elusive, but because physicists have been looking for the wrong thing.

Let's break down the illusion:

It doesn't emit light:

- Because it's not excited
- It doesn't ripple unless disturbed
- So it remains invisible to telescopes

It doesn't resist motion:

- Because it's not condensed
- It offers ripple continuity, not mechanical friction
- So it doesn't behave like gas, dust, or plasma

But it maintains ripple alignment:

- Across galactic scales
- Between massive structures
- And within gravitationally weak regions

So while dark matter is imagined as a sea of undetectable particles, what's really happening is that stars are floating in a coherent ripple web—a field that holds their motion in check without exerting gravitational force.

Galaxy Cohesion Without Mass

This explains why galaxies:

- Don't tear apart despite fast outer star rotation
- Don't require extra gravitational pull
- Remain stable over billions of years

It also explains why dark matter doesn't clump like matter, why it doesn't form disks, and why it fails to interact in expected ways.

Because it isn't matter. It's structure.

Gravitational Lensing Anomalies

When light bends around massive clusters more than expected, physicists assume "extra mass" must be bending spacetime.

But in UED, the bending isn't from mass—it's from ripple phase delay through Stable SVE.

Light isn't responding to mass, but to field topology. Stable SVE causes ripple slowdown or deflection—not through gravity, but through phase-based resistance in field dynamics.

Why Particle Detectors Always Fail

Billions have been spent trying to detect dark matter particles—WIMPs, axions, supersymmetry candidates. All attempts fail.

Why?

Because Stable SVE is:

- Not particulate
- Not interactive
- Not mass-based

It is a non-condensed field. No particle physics experiment will find it, because they are all built to detect condensed ripple events, not stable field continuity.

UED's Conclusion

The mystery of dark matter vanishes once you stop demanding mass as the explanation for motion.

Stable SVE explains:

- Galactic structure
- Lensing behavior
- Detection failure

All without resorting to imaginary particles.

Dark matter is not hidden mass. It is undisturbed structure. It's not what's missing—it's what's quietly present, doing its job without drawing attention.

In the next section, we show how this insight doesn't just solve one problem—it rebuilds the entire galactic model. We move from ghost-hunting to clarity—replacing 40 years of speculation with a simple, ripple-aware logic.

From Mirage to Model: UED's Galactic Clarity

"What you called 'missing mass' was a ripple misunderstanding."

With the introduction of Stable Structured Vacuum Energy (SVE), the galaxy stops being a gravitational mystery and becomes a ripple-balanced structure. Under Universal Energy Dynamics (UED), we no longer need to invent ghost particles, simulate invisible halos, or build billion-dollar detectors in the hope of finding a hypothetical mass that never existed.

The entire model shifts—from chasing matter to understanding field conditions.

No Exotic Particles. No Dark Halos. No Gaps to Fill.

Mainstream physics has spent decades trying to build around the "dark matter problem":

- Particle physics has proposed unverified entities like WIMPs, axions, sterile neutrinos, and supersymmetric candidates.
- Astrophysics has modeled enormous spherical "dark matter halos" enveloping galaxies to match observations with Newtonian or relativistic expectations.
- Cosmology has accepted that 80–85% of all matter in the universe is "dark," despite having zero direct detection to date.

And yet, none of these theories have closed the gap.

Why? Because the gap was never about matter—it was about interpretation.

Galaxy Dynamics through Ripple Mechanics

UED replaces this speculative framework with a ripple-based, testable, and consequence-driven understanding of galactic behavior. It sees galaxies not as gravitationally bound mass-collections, but as coherently structured ripple environments.

1. Ripple Anchoring

Stars are not held by invisible mass; they are anchored within ripple-aligned zones of SVE. This anchoring allows stars to move at high velocities without being ejected—because they are in phase with the ripple framework surrounding them.

These ripple structures don't break or stretch—they realign locally to maintain the pattern. The result is dynamic cohesion without mass-based force.

2. ATT: Alignment Threshold Time Across Cosmic Distances

ATT is the time-sensitive window that allows ripple exchange or synchronization between two zones.

- In galactic systems, massive bodies influence local ripple configuration.
- ATT ensures that distant stars are still functionally coupled to the central structure, not through gravity, but through ripple coherence.
- This coherence doesn't fade with distance the way gravitational force does—it holds as long as ATT conditions are met.

This explains why outer stars maintain high speeds—they are still in ripple alignment, even if they're far from the core.

3. Entropy Field Gradients

Every ripple structure in UED is subject to entropy flow—the movement of ripple tension toward equilibrium.

In galaxies:

- Stars are not just orbiting—they are participating in entropy regulation.
- The ripple field between stars flows entropy outward, adjusting ripple geometry.
- This keeps the system rotationally stable, without any need for extra matter to explain inertia.

Rotation, in this view, is not maintained by mass—it is maintained by entropy symmetry in the ripple field.

From Ghost Story to Grounded Physics

This new model offers clarity, not complexity.

- No more dark halos.
- No more undetectable particles.
- No more energy budgets inflated with hypothetical entities.

Instead, UED explains galaxy behavior using:

- Structured vacuum geometry
- Ripple alignment conditions
- Real-time entropy distribution

This is not a patch—it's a replacement.

The "dark matter problem" dissolves once you understand what ripple anchoring really is. The mass was never missing—the method was.

Final Insight

Dark matter was never the solution. It was a symptom of a physics model built on ripple consequences, not ripple causes.

UED removes the illusion. It shows that the universe doesn't need to hide 80% of its matter.

It only needs us to finally listen to the field—not the echo.

What you called 'missing mass' was a ripple misunderstanding.

In the next chapter, we confront the other half of the "darkness" myth—dark energy. And again, what's interpreted as repulsive force turns out to be ripple behavior misread through entropy gradients.

Chapter 4: The Phantom Force Called Dark Energy

Why the universe's so-called acceleration is not force-based, but ripple-based. No push. No repulsion. Just divergence.

In 1998, the universe broke its script.

Two independent research teams, studying distant Type Ia supernovae, found something no one expected: the light from these ancient explosions was dimmer—and redder—than predicted. According to every cosmological model available, the universe's expansion should have been slowing down, gradually succumbing to gravity's pull.

But it wasn't. It was speeding up.

Or so it seemed.

This single anomaly cracked open the foundation of modern cosmology. It didn't just suggest something new—it demanded something unnatural. A force unlike anything ever observed. A mysterious, invisible, all-pervasive repulsion. A pressure woven into the fabric of space itself.

And so, like its gravitational twin dark matter, this new actor was summoned into existence by theory before it was ever seen.

They called it dark energy.

But what is dark energy? To this day, no one knows. It's not a field, not a wave, not a particle. It has no origin, no interaction, no behavior—except one: it accelerates the expansion of space.

Or so we say.

But what if that acceleration isn't real? What if nothing is being pushed apart at all?

What if what we call "expansion" is just ripple divergence—entropy flowing naturally through a structured vacuum field?

What if redshift doesn't mean speed—it means ripple delay?

What if dark energy is not a thing—but a symptom of how we misread the field between us and the light?

Maybe the universe is not accelerating outward. Maybe it's just unwinding. Maybe we're not witnessing a push—but a gentle return to stillness.

This chapter will dismantle the myth of dark energy—not by denying the data, but by offering a better explanation: no push, no repulsion—just divergence.

Let's decode the phantom.

The Supernova Surprise

"When the light stretched too far, we blamed the fabric, not the thread."

In the late 1990s, a quiet revolution erupted in the night sky.

Two independent teams of astrophysicists—the Supernova Cosmology Project led by Saul Perlmutter, and the High-Z Supernova Search Team led by Brian Schmidt and Adam Riess—set out to measure the expansion rate of the universe by observing distant Type Ia supernovae. These stellar explosions are considered "standard candles" due to their consistent brightness, allowing astronomers to calculate both distance and relative velocity based on observed redshift.

But what they found changed everything.

The Unexpected Redshift

The supernovae were fainter and redder than expected—not because they were farther away than anticipated, but because their light had stretched more than it should have by the time it reached Earth.

This meant the expansion of the universe wasn't slowing down under the influence of gravity. It was accelerating.

This defied every expectation of classical and relativistic physics. Gravity—despite all its weaknesses on cosmic scales—was still expected to provide a long-term braking effect on expansion, especially given the matter content of the universe.

But here was hard evidence that the opposite was occurring.

The Shaking of the Standard Model

Faced with this empirical contradiction, the scientific community stood at a crossroads. They could:

1. Re-express the nature of the vacuum.
2. Revisit gravity as an incomplete theory at cosmic scales.
3. Consider the possibility that redshift did not correspond directly to velocity or expansion.
4. Or—create a new force.

They chose the fourth.

To preserve the existing framework, cosmologists introduced a repulsive force embedded within the vacuum itself—an unknown form of energy that pushes galaxies apart faster over time.

They named it Dark Energy.

Not because it had been observed, but because the equations needed something to explain the redshift anomaly.

Just like dark matter, dark energy was not a discovery. It was a patch—a placeholder to stabilize a model collapsing under the weight of unexpected data.

The Return of Λ — The Cosmological Constant

Ironically, this "new" force had an old name.

In 1917, Einstein introduced the cosmological constant (Λ) to keep the universe static, in line with the prevailing belief that the cosmos neither expanded nor contracted. When Hubble later proved that the universe was expanding, Einstein famously called Λ his "biggest blunder" and discarded it.

But in 1998, Λ made a dramatic comeback.

This time, it wasn't used to hold the universe still—it was used to accelerate it outward. Cosmologists rebranded the constant as a form of vacuum energy density—a sort of anti-gravity sewn into the fabric of spacetime.

Thus, Λ was reborn—not as a cosmological curiosity, but as the dominant term in the energy budget of the universe, allegedly responsible for nearly 70% of all cosmic behavior.

Yet this rebirth carried an unspoken cost.

The Philosophical Blind Spot

Rather than revisiting the foundational assumptions about redshift, expansion, and the behavior of light through vast stretches of structured vacuum, science chose to double down on a framework built on ripple consequences.

Redshift was assumed to indicate velocity. Velocity implied spatial separation. And increasing redshift meant acceleration.

But what if redshift meant something else? What if light stretched not because space expanded, but because the ripple conditions between source and observer changed?

What if this was not a runaway universe—but a ripple divergence phenomenon?

When the light stretched too far, we blamed the fabric—not the thread that carried it.

In the next section, we'll examine this fabric more closely. Because dark energy, like dark matter, is not a discovery. It's a misinterpretation—one rooted in a failure to understand how ripple behaves across a still but structured vacuum field.

Let's now decode why this mysterious "force" isn't a force at all—but a field realigning through ripple entropy.

The Force That Wasn't There

The most powerful force in the cosmos... never showed up.

After the supernova shock of 1998, cosmology faced an existential question:

What could possibly be pushing the universe apart with increasing speed?

The answer, according to mainstream physics, was Dark Energy—a mysterious force unlike any other in the known universe.

But unlike gravity, electromagnetism, or the nuclear forces, dark energy does not behave like a force in any testable or observable sense. It doesn't interact. It doesn't fluctuate. It doesn't respond.

It simply exists—in theory.

A Force in Name Only

Let's examine the attributes of this so-called "force":

It doesn't interact

Dark energy exerts no measurable pressure on matter. It cannot be shielded, deflected, or blocked. No experiment has recorded it altering the behavior of particles, fields, or even spacetime in any controlled way.

It doesn't cluster

All known forces tend to concentrate in high-energy or high-mass regions. Gravity pulls mass together. Electromagnetism forms charged structures. Even the strong force binds atomic nuclei.

But dark energy? It's completely uniform. It doesn't gather, accumulate, or respond to environmental changes. It doesn't do anything—it just remains.

It doesn't obey field dynamics

There is no field equation for dark energy that behaves like a conventional interaction. No Lagrangian. No mediating boson. No symmetry breaking. It cannot be modeled as a real physical interaction in the way the other fundamental forces can.

No Carrier, No Interaction, No Explanation

Dark energy has no particle associated with it—no "dark boson" or force-carrier.

There's no known interaction channel. No direct signal. No resonance. No quantization. It is completely detached from all other physical frameworks.

And yet, cosmology boldly claims that this ghostly non-entity drives over 70% of the universe's dynamics.

That's not physics. That's placeholder metaphysics.

We named a force because the data didn't match the model. We declared that force dominant—even though it has never acted.

The UED Critique: A Misreading, Not a Mystery

Universal Energy Dynamics offers a sharp critique—not by denying the redshift data, but by exposing the misinterpretation of its meaning.

In UED, there is no dark energy force—because there is no force needed to explain cosmic divergence.

Here's what UED sees instead:

- The so-called "acceleration" is not motion—it's ripple divergence across structured vacuum.
- The redshift is not caused by recession—it's caused by ripple entropy flow, altering how light stretches through vast, ripple-thinning fields.
- There's no mysterious outward push—just progressive ripple dilution in a system returning toward equilibrium.

In other words:

What we mistook for repulsion was simply dissipation. What we labeled as dark energy was just entropy expressing itself in structured silence.

Why the Fabric Misled Us

Physicists looked at stretched light and said, "space must be expanding faster."

But UED says: the field between source and observer was relaxing, losing ripple density over time. This made light appear stretched—not because of movement—but because it traveled through a thinning medium.

It's not spacetime inflating. It's ripple coherence dissolving. No force is pushing—it's simply less holding it back.

This is not expansion by propulsion—it's divergence by passive field decay.

Summary: The Force That Never Was

Claim (Mainstream)	UED Interpretation
Universe is accelerating outward	Ripple fields are relaxing outward
Caused by a repulsive vacuum force	Caused by entropy-driven ripple divergence
No carrier, no interaction, no behavior	Because it's not a force—it's a misread
70% of universe explained by dark energy	70% of universe is ripple stillness being rebalanced

Dark energy isn't mysterious. It's unnecessary.

What we perceived as a force was the absence of resistance. A silent exhale mistaken for a cosmic shove.

In the next section, we dive deeper into what's really happening: the mechanics of ripple divergence and entropy gradients, and how they reshape our understanding of cosmic behavior—without invoking imaginary forces.

Ripple Divergence and Entropy Gradient

"This isn't a universe accelerating. It's a field exhaling."

The dominant narrative in modern cosmology claims that the universe is not only expanding, but doing so at an ever-increasing rate. The metaphor often used is that of a balloon being inflated—galaxies drifting farther apart as space itself stretches between them.

But what if the balloon isn't expanding at all? What if it's simply relaxing?

Universal Energy Dynamics (UED) replaces the picture of a ballooning cosmos with one far more elegant—and far more grounded in energetic consequence.

The universe isn't accelerating outward. The field is exhaling. Entropy is flowing, and ripple structures are realigning into lower-tension states.

What we interpret as "galaxies receding" is not motion through space, nor space expanding—it is the appearance of divergence caused by ripple behavior through a structured, but relaxing, vacuum field.

Expansion = Ripple Entropy Flow, Not Space Stretching

In UED, space is not a geometric fabric that expands—it is Structured Vacuum Energy (SVE), a still, energetic field capable of holding and transmitting ripples. "Expansion" is not the field stretching—it's the dissipation of ripple coherence within that field.

Entropy is the driver.

As ripple fields spread out, the tension between them decreases, and the system seeks equilibrium. The energy is not accelerating outward—it is diffusing across the field's available entropy gradient.

This results in:

- Fewer ripple alignments
- Lower ripple density

- Slower energy exchanges

The outcome looks like acceleration. But it is ripple divergence, not spatial expansion.

Why Distant Galaxies Appear to Recede

In ripple mechanics, the motion of galaxies is not absolute—it is a byproduct of how ripple signals travel through SVE.

Their apparent velocity increases with distance not because of physical speed, but because of:

1. Loss of Ripple Density

As you move further from any concentrated energy zone (like a galaxy), the ripple density decreases. That is, the number of coherent ripple lines per unit region drops.

Lower ripple density means:

- Weaker coupling with other zones
- Greater delay in interaction
- Reduced resonance with the observer's reference frame

This results in signal stretching, which is interpreted as redshift.

2. Increased Ripple Delay (RDI)

Ripple Delay Index (RDI) measures how long a ripple signal takes to propagate between points across SVE. As entropy increases, ripple signals:

- Take longer to arrive
- Arrive with less intensity
- Are stretched in both time and frequency

So when light from a distant supernova arrives "redder," it's not because the galaxy is flying away—It's because the ripple carrying the light has slowed and stretched through thinning ripple fields.

3. Decreased ATT (Alignment Threshold Time)

ATT is the temporal condition needed for two regions of SVE to couple ripple information effectively.

Over vast cosmic distances:

- Ripple fields drift out of alignment
- ATT conditions are rarely met
- Fewer high-fidelity exchanges occur

The result? Observers detect less energetic coherence, interpreting this as acceleration—when in reality, it's just failure of field alignment over distance.

Not Repulsion—Just Redistribution

Mainstream physics interprets the redshift anomaly as a sign of outward force—a repulsive energy embedded in the vacuum. But under UED, there is no need for force at all.

The system is simply:

- Releasing ripple tension
- Redistributing energy over wider field geometries
- Allowing entropy to equalize phase disparity

This isn't expansion due to pressure. It's natural ripple divergence due to field relaxation.

Galaxies aren't being pushed apart. They're slipping out of ripple sync with each other—falling into silence, not motion.

Rewriting the Cosmic Behavior

Mainstream Interpretation	UED Interpretation
Accelerating expansion	Ripple divergence across low-density SVE zones
Caused by a repulsive force	Caused by entropy-driven ripple thinning
Increasing galactic separation	Increasing loss of ripple coherence
Cosmological constant (Λ) required	Λ is a misread of entropy gradient dynamics

Closing Perspective

This ripple-based interpretation transforms our understanding of cosmic behavior:

The universe is not expanding like a balloon being inflated—it's a ripple system winding down, letting go of tension, returning to coherence.

There is no "dark force" pushing galaxies apart. There is only ripple entropy flowing through a structured field too subtle to be seen—but too perfect to fail.

What we called acceleration was just relaxation misunderstood.

In the next section, we reframe redshift itself—not as a velocity measure, but as a ripple signal distortion over cosmic distances. It's time to see the color of light as a consequence of field behavior, not motion.

Rethinking Redshift with Ripple Logic

Light didn't race away—it just took the long way home.

Redshift is the cornerstone of modern cosmology's understanding of the universe's expansion. It is the stretching of light's wavelength as it travels across space. For over a century, this phenomenon has been interpreted in two dominant ways:

1. **Doppler Shift** – If the source of light is moving away, the wavelength gets stretched. Just like a train's whistle sounds

lower as it recedes, galaxies "sound" redder as they move from us.

2. **Spacetime Expansion** – Under general relativity, light traveling through an expanding spacetime gets stretched along with it. This is not motion through space, but space itself growing, pulling the wavelength longer as it propagates.

Both interpretations rely on a motion-based worldview—one assumes that objects are moving, the other assumes that the medium itself is.

But what if neither is true?

What if redshift is not a measure of motion—but a consequence of ripple behavior through a dynamic, entropy-laden field?

This is where Universal Energy Dynamics (UED) delivers a profound shift in cosmological interpretation.

Not Doppler. Not Geometry. Just Ripple Delay.

In UED, light is not a photon particle traveling through space, nor a wave on a fictitious spacetime fabric. It is a ripple alignment event within Structured Vacuum Energy (SVE)—a field that carries ripple signals through resonance, not motion.

And when light appears redshifted, it isn't because:

- The galaxy is moving away (Doppler), or
- Space is stretching (relativistic expansion),

It's because the ripple pathway it traveled through has changed in entropy and coherence.

In UED:

- The vacuum field is not static.
- Its ripple density, tension, and alignment capacity evolve with time and distance.

- Over cosmic scales, the entropy of the field increases—causing ripple signals to slow, stretch, and scatter.

The redshift is not a mark of recession. It is a ripple distortion caused by a relaxed field.

Why Ripple Signals Stretch in a Relaxed Field

As entropy rises across a vast region of SVE:

- Ripple signals lose intensity and resolution.
- Their alignment threshold (ATT) with the observer decreases.
- They take longer to arrive—not because of distance, but because of delay within the medium.

This delay is not a slowing of light—it is a prolongation of ripple coupling across zones that no longer support high coherence.

What reaches the observer is:

- A low-tension ripple, arriving late,
- With a stretched wavelength,
- Causing the signature we read as redshift.

But in truth, the source never moved. The medium simply stopped helping the signal along.

Ripple Thinning Across Entropy Gradients

UED introduces the concept of ripple thinning: the weakening of ripple coherence due to the gradient of entropy across a field.

When energy moves through:

- A dense, well-aligned SVE field: signals stay intact.
- A thin, entropy-saturated field: signals degrade, lose energy, and stretch.

The further the signal travels, the thinner the field becomes—not because it expands, but because it energetically relaxes.

The redshift, then, is not a record of how fast something moved away. It's a record of how the field changed between the event and the observer.

The Consequences of Misreading Redshift

The entire notion of an accelerating universe collapses when redshift is understood this way.

- There is no need to invent anti-gravity.
- No need to patch the equations with dark energy.
- No need to explain why distant galaxies appear to move faster.

They aren't. They never were.

Their light is just passing through increasingly incoherent ripple fields, where:

- ATT is harder to achieve,
- RDI (Ripple Delay Index) increases,
- And ripple thinning makes every signal look like it came from a faster, further source.

Redshift as Ripple History

UED redefines redshift as a historical record of ripple environment quality—not an indicator of object velocity.

Conventional View	UED Interpretation
Redshift = velocity of recession	Redshift = ripple delay and entropy divergence
More redshift = faster expansion	More redshift = weaker field coherence
Redshift caused by spacetime geometry	Redshift caused by ripple-path deformation
Requires cosmic inflation and Λ	Requires only entropy gradient across SVE

Final Thought

We've been treating redshift as a motion blur. In truth, it's a ripple memory—a stretched echo from a field that has since exhaled.

The light didn't flee—it simply aged across a thinning silence. The signal didn't accelerate—it drifted through surrender. The field didn't push—it let go.

No anti-gravity. No exotic energy. Just ripple behavior through entropy-aware vacuum.

In the next section, we resolve the myth entirely. We ask: If dark energy isn't a force, what is it really? And the answer lies not in mathematics, but in energetic stillness misunderstood.

What Dark Energy Really Is (UED View)

"Dark energy is not the engine of expansion—it's the echo of a universe returning to stillness."

The universe is not expanding *because* of a force. It only appears to be, because we're trying to measure the relaxation of a cosmic field with tools designed to read tension.

In the UED framework, dark energy does not exist as a physical entity, particle, or repulsive substance. It is not a pressure in the vacuum, nor a mystery force woven into spacetime.

So what is it?

It is the consequence of ripple field divergence over vast stretches of Structured Vacuum Energy (SVE)—a structured, still foundation that slowly unwinds energetic tension through entropy.

Dark energy, in this view, is not what's causing expansion. It is the symptom of an expanding misunderstanding.

No Substance. No Force. Just Divergence.

Let's be clear: Dark energy has no mass, no charge, no known field carrier, no quantum interaction, and no observed fluctuation—yet it supposedly drives 70% of the universe's dynamics.

This is not a discovery. It is a placeholder for our confusion.

UED clarifies this by showing that what we call "dark energy" is simply:

- A ripple field diverging across cosmological scales
- Releasing tension into entropy
- Without compression or feedback

This behavior is not driven by any "anti-gravity" force. Rather, it's a directional entropy relief—the cosmos exhaling back into balance.

Expansion Is a Perception Artifact

The so-called accelerating expansion is not an actual acceleration. It's a misread of ripple dynamics caused by:

1. Observer Ripple Frame Misalignment

Observers (like us) measure the universe using light and motion—two ripple-based effects. But our tools are phase-locked to our local SVE conditions. As we try to interpret signals from distant regions:

- The ripple frame has shifted
- The entropy of the field has changed
- And the temporal rhythm (ATT) between us and the signal source has decayed

The result? We perceive signal stretch and interpret it as object movement—when in fact, it's field misalignment.

We are reading a story the universe already stopped telling—through tools still echoing the past.

2. Measurement Across Incoherent SVE Regions

As ripple waves pass through increasingly relaxed, low-density vacuum zones:

- They lose coherence
- They delay in arrival

- They undergo entropy-induced deformation

This creates the illusion of acceleration, because the information is arriving slower and longer—but not due to recession or repulsion.

The universe is not speeding up. Our perception is falling out of sync with its stillness.

Ripple-Based Reinterpretation of Λ (Cosmological Constant)

Einstein's cosmological constant (Λ) was originally introduced to keep the universe static. It was later resurrected to explain the opposite—expansion.

In UED, Λ is neither needed nor meaningful. The behavior it describes is not a force or field—but a natural consequence of ripple field divergence:

Λ as "dark energy"	UED reinterpretation
Adds pressure to expand space	No space expansion—only ripple spread
Opposes gravity	No opposition—only entropy relaxation
Uniform, isotropic force	Uniform entropy flow across SVE zones
Requires unmeasurable energy field	Requires only ripple behavior and field memory

Thus, what Λ tries to explain geometrically, UED explains energetically.

Dark Energy = Entropy's Silent Work

Let's call it what it is:

- Not repulsion
- Not expansion
- Not pressure

It's the relaxation of ripple structure—an exhale from a system that was once intensely aligned, now returning toward equilibrium.

When the ripple tension diminishes:

- No further alignment happens
- Light fades and stretches
- Signals slow

And we, still tied to ripple perception, think the source moved.

Final Reflection

Dark energy is not the fuel of expansion. It is the illusion left behind when the ripple field diverges and lets go.

It's what we see when we try to observe stillness using the lens of motion.

The universe is not racing away from us. It's retreating into rest. Dark energy is not a cause—it's the absence of cause still echoing through our equations.

No force. No mystery. Just ripple fading into calm.

Chapter 5: Gravity Without Mass?

Redefining attraction as ripple gradient response—not mass curvature. A complete departure from Einstein's fabric.

What if gravity was never a force? What if space was never curved?

And what if we've spent centuries chasing the wrong shadow—trying to measure a pull, when all we ever needed to notice... was a slope?

From Newton's invisible tug to Einstein's elegant curvature, gravity has worn many masks. But in every version, mass was the storyteller. The lead actor. The cause. The origin. No mass, no gravity—that was the deal.

Until galaxies rotated too fast. Until light bent through empty fields. Until invisible "dark matter" had to be invented... just to make the equations breathe.

We've been told that gravity is what mass does to space. UED says: It's what the ripple field does to itself.

There is no curvature, only imbalance. No invisible pull, only a realignment of energy. No mysterious force—just the universe, correcting its own tension.

"You thought you were falling. But you were only sliding down a ripple."

Let's now begin the journey that breaks gravity away from mass—and lets the universe exhale.

Gravity—A Force Invented, Then Bent

"When we couldn't touch the cause, we modeled its shadow."

Gravity—the grandest mystery cloaked in certainty.

It began with Isaac Newton, who gave the world a force that worked flawlessly, but only mathematically. Newton's law of universal gravitation—

$$F = G\frac{m_1 m_2}{r^2}$$

—was revolutionary. Yet, Newton himself admitted he didn't know how it worked. He called it *"action at a distance"*—a force that magically operated between bodies, with no medium, no contact, and no explanation. The apple fell. The Moon orbited. The numbers added up. But gravity, in his view, was a mysterious pull, justified only by its effects.

Then came Einstein, who replaced the invisible pull with a geometric illusion. In General Relativity, gravity wasn't a force—it was the result of mass bending spacetime. Matter told space how to curve; curved space told matter how to move. Elegant, poetic... and again, entirely dependent on mass being present.

Whether it's Newton's pull or Einstein's curve—both frameworks treat mass as the central agent. No mass, no gravitational influence. The equations breathe only when mass enters the room.

But reality has quietly resisted this narrative.

In the outskirts of galaxies, where luminous mass fades, stars still rotate at bewildering speeds. In intergalactic voids, light bends around seemingly empty space. In clusters, lensing patterns emerge that cannot be explained by the visible matter alone. These are not mathematical quirks—they are repeated, observable phenomena. Something is shaping motion—but that something has no mass signature.

And so, rather than question the premise, physics added layers:

- **Dark Matter:** to explain massless gravity.

- **Dark Energy:** to explain massless expansion.
- **Gravitons:** to explain quantum pull.
- **Modified Newtonian Dynamics (MOND):** to salvage the old framework without rethinking its base.

All these attempts stem from one refusal—to let go of mass as the origin point.

When galaxies refused to slow down, we didn't question the math. We invented something invisible to hold up the illusion.

This is the gravitational myth: a centuries-old habit of explaining motion by inventing mass behind it. But what if there was no "behind"? What if gravity was never about pulling, but about settling—an alignment across ripple fields, triggered not by mass, but by imbalance in structure?

As we proceed, we will see how UED redefines gravity—not as attraction, not as curvature, but as a ripple gradient. A consequence of field topology trying to regain equilibrium. No force. No bend. Just a slope... invisible to the eye, but felt by all motion.

And for the first time, we will ask: What happens when mass isn't the cause—but the consequence?

The Ripple Gradient Hypothesis

"Mass doesn't pull. It realigns what surrounds it—and everything follows the slope."

If mass does not pull, and space does not bend, then what exactly is gravity?

Universal Energy Dynamics (UED) offers an entirely new lens:

Gravity is not a force, but a ripple gradient. Not attraction. Not curvature. But a field response to localized energetic imbalance—a slope formed within a structured, ripple-sensitive vacuum.

Field First, Not Force First

In UED, Structured Vacuum Energy (SVE) is not empty—it is a *ripple-ready field*. Still by nature, yet capable of disturbance. When a ripple begins—say, through the condensation of energy into mass—it doesn't just create an object. It alters the alignment of the field around it.

This disturbance is not a gravitational force in the Newtonian sense, nor a geometric curvature in Einstein's sense. Instead, it forms a gradient of ripple tension—a *sloping terrain* across the SVE fabric.

Mass as a Ripple Disturbance

Mass is nothing more than condensed ripple complexity. It is not a source of force, but rather a modifier of alignment. Like placing a stone in a still pond—not to generate waves, but to reshape how ripples will travel next.

That stone creates a gradient—a zone around it where the ripple field's behavior changes. Not because the stone exerts force, but because the field responds to its presence.

Now place another object—a neighboring ripple system—within that same pond. It will not be pulled. It will simply descend along the slope of lesser resistance. This is not gravity. It is ripple alignment in action.

The Role of ATT: Timing the Descent

This realignment process is governed by a key UED parameter:

ATT – Alignment Threshold Time. It defines the *delay* required for a field region to adjust to an external ripple's influence.

- In strong gradients, ATT is short: alignment is quick—appearing as stronger gravity.
- In subtle gradients, ATT is longer: alignment is slower—weaker gravity.

Thus, what we perceive as acceleration under gravity is not due to a force being applied. It's the timing of a natural field realignment. Bodies do not fall—they settle into the least resistant ripple path.

Beyond Local Gravity: Field-Wide Implications

This model explains why:

- Gravity works even in vacuum
- Gravitational lensing happens in non-massive zones
- Gravitational anomalies occur where mass is undetectable

It also removes the need for a carrier (like gravitons), or for infinite range forces. Ripple gradients form, extend, and fade based on entropy distribution and field rigidity.

There is no central puller in the system—only distributed slopes that guide motion. The more condensed a ripple (mass), the steeper the gradient. But even without mass, stable SVE can hold shape and guide motion across galactic scales.

"Mass doesn't pull. It realigns what surrounds it—and everything else rolls into place."

This is not a tweak to Newton or Einstein. This is a departure from the need for pull, bend, or force. This is gravity as gradient—pure, ripple-responsive, and fully field-based.

Why Mass Is Not Required

"Gravity isn't what mass does—it's what the field remembers."

In classical physics, mass is the gatekeeper of gravity. No mass, no gravitational pull. But the universe has been whispering a different truth—one that mass-based frameworks consistently ignore.

Because gravity shows up where no mass can be found.

Gravity in the Void

In vast interstellar regions, stars remain in cohesive orbits even where visible matter thins out. In cosmic voids, gravitational lensing distorts light paths—despite no mass being detected. In regions mapped as "dark matter halos", we observe coherent gravitational influence—without particles, photons, or any direct interaction.

And yet, the equations insist:

"Something *must* be there. Some hidden mass. Some exotic particle."

But what if the field itself is the something?

The Field Has Memory

Under Universal Energy Dynamics (UED), the Structured Vacuum Energy (SVE) is not just passive backdrop. It is field with memory—with ripple history, phase alignment, and intrinsic stability zones.

Where mass once disturbed the field, or where a ripple pattern became stable, that configuration persists. And because the field is continuous, this persistent structure creates gradients—not due to mass, but due to ripple history.

This is why gravitational influence exists even when:

- Mass is gone
- No light is emitted
- No known force is interacting

It's not the object that causes gravity. It's the lasting shape it leaves in the field.

Stable SVE: The Silent Scaffold

Stable SVE zones can:

- Anchor motion (by creating ripple descent paths)
- Bend light (through phase interference, not curvature)
- Hold systems together (like galactic halos)

All without emitting, resisting, or possessing mass.

These SVE structures may appear "dark" to light-based instruments because they don't ripple back. They exist in perfect alignment, and thus offer no response to electromagnetic probing.

This is why dark matter has remained undetectable—not because it's hidden, but because we're looking for ripples, not structure.

The Mass Misreading

We've mistaken ripple gradients for massive pull.

What we call "gravitational mass" is simply the visible proxy for what's actually a field condition. A localized ripple configuration that affects motion, not because it attracts, but because it provides the path of least resistance.

Much like air pressure moves wind—not because the air is pushing, but because pressure gradients exist—gravity too is simply alignment pressure in the ripple field.

"Mass is visible. Memory is not. And gravity listens to memory."

We no longer need to chase exotic particles, weakly interacting masses, or hidden substances. We just need to read the field.

And UED gives us the language.

Ripple Mechanics vs. Curved Geometry

"You don't bend space to move a planet. You align energy to show it where to settle."

"Mass tells space how to curve, space tells mass how to move."

This famous summary of General Relativity by John Wheeler captures Einstein's vision of gravity. Elegant, geometric—and fundamentally misaligned.

Because if space must bend for mass to move, then we've anchored physics in a metaphor. A fabric that no one can touch. A curvature that no one can measure directly. And every ripple—light, energy, gravity—is assumed to follow these invisible folds.

UED steps in and says:

Stop bending space. Start reading the ripples.

From Curvature to Alignment

In Universal Energy Dynamics (UED), mass is not a "thing" that deforms a canvas. It is a condensed ripple formation—a localized disruption in an otherwise still and structured field (SVE). This disruption alters how ripples travel through that field, creating alignment gradients—zones of higher or lower field tension.

Objects, energy, and light do not "fall" or "curve" because space is bent. They simply realign along the ripple slope—like water flowing down an energetic hill.

The movement is not caused by gravity. It's caused by ripple path optimization.

Orbits as Ripple Feedback Loops

In this model, orbits emerge not from central attraction, but from balanced alignment.

A planet doesn't fall toward the sun because of curved spacetime. It moves within a dynamic feedback loop—where its ripple field interacts with the solar ripple gradient, settling into a harmonious orbit defined by:

- Local ATT (Alignment Threshold Time)
- Ripple tension zones
- Entropic resistance and release patterns

When ATT stabilizes, so does the orbit—not due to inertia, but due to equilibrium in ripple feedback.

Lensing Without Curving

Einstein predicted gravitational lensing because mass supposedly curves space and bends light paths.

UED offers an even simpler explanation: Ripple speed is field-dependent. Just like sound travels slower through fog, ripple propagation slows in certain SVE conditions.

Thus, light bends not because geometry is distorted, but because:

- Ripple speed varies in structured vacuum
- Light deflects along entropy gradients
- The field's ripple density reshapes wavefront coherence

What appears to us as lensing is simply ripple diffraction across misaligned SVE zones.

Gravitational Waves Reimagined

Einstein's prediction of gravitational waves was confirmed in 2015—ripples in spacetime generated by massive collisions. But what if those waves weren't in space, but in the ripple field?

UED reinterprets them as:

- Entropic ripple echoes
- Transient ATT distortions
- Rebalancing surges in the SVE field

No spacetime wobble. Just field compression and realignment—like ripples through gelatin, not curves through fabric.

These echoes do not need curved geometry to explain their propagation. They only need ripple mechanics responding to energetic events.

Implications for the Future

If ripple dynamics governs:

- Orbits
- Light deflection
- Energy transfer
- Cosmic structure

Then we must abandon the geometry-centric narrative. It has served its time—but now restricts precision and understanding.

Ripple mechanics is not an alternative to Einstein. It is an evolution beyond him.

It doesn't erase the results—it recontextualizes them with foundational clarity.

"Space doesn't curve. Ripples align. And in that alignment, reality moves."

The Equation Reimagined

"We never needed gravity to be mysterious. We just needed to stop forcing it to be mass-bound."

For centuries, gravity has worn a rigid mathematical cloak:

$$F = G\frac{m_1 m_2}{r^2}$$

Newton gave us this iconic formulation—a law of attraction, elegant and precise. Yet its elegance came at a cost: dependence on mass, a variable often inferred rather than observed.

Later, Einstein added geometry, replacing force with curvature—yet still dependent on mass-energy tensors to explain how the "fabric of spacetime" bends.

In both cases, mass is assumed as the origin of gravitational behavior.

But if mass is only the *condensed ripple effect* in the structured vacuum, then we must rebuild the entire gravitational logic. That's where UED offers a new equation—stripped of mass, rooted in ripple behavior.

The Ripple Gradient Equation

$$R_g = \frac{\Delta \rho_r}{\Delta d}$$

Where:

- R_g: Gravitational response or ripple-induced alignment pressure
- $\Delta \rho_r$: Ripple density gradient (difference in ripple memory/

tension between two points)

- Δd: Field distance or ripple path length through Structured Vacuum Energy (SVE)

This isn't a force. It's not even an acceleration. It's a tendency—a ripple gradient seeking equilibrium, expressing itself through energy realignment.

Instead of "mass attracts mass," we now have:

"Ripple density flows from distortion to equilibrium."

It is a gradient velocity, governed by field tension and ripple delay, not an applied force acting at a distance.

Implications for Prediction and Observation

Let's break down where this reformulated equation becomes more useful than Newtonian gravity—especially when no mass is visibly present:

1. Galactic Motion

- Rotation curves no longer require invisible mass.
- The ripple gradient across SVE explains consistent orbital speeds as a result of field-based anchoring.

2. Gravitational Lensing

- Light bends not due to curved geometry but due to changes in ripple density—slowing or diverting path like light in varying optical media.
- $\Delta\rho_r$ becomes a lensing predictor.

3. Void Interactions

- Where no mass is found (dark regions, cosmic filaments), the ripple equation still predicts gravitational-like interactions, because stable SVE stores ripple tension and propagates alignment gradients.

4. Planetary and Satellite Behavior

- Can explain perturbations in orbital paths due to ripple interference—such as solar flares, dense ripple events, or gravitational anomalies, without invoking hidden mass or exotic forces.

Replacing Force with Flow

In this model, acceleration is not the default result. Rather, it's apparent velocity due to realignment speed in the ripple field. The concept of *gravitational potential* gets replaced by:

Ripple entropy topology—the slope of ripple memory density that governs motion.

There's no longer a need for "gravitational attraction." Instead, we observe field response—energy returning to rest.

The universe doesn't attract. It rebalances.

Legacy Equations Revisited

- Newton's law is useful for Earth-bound mechanics, but fails at cosmic scale.
- Einstein's tensors add accuracy but hide cause in complex spacetime abstractions.
- UED offers clarity: ripples are not effects of curvature—they are the primary field language.

Thus, UED reimagines gravitational modeling as field-centric, not mass-centric.

"Gravity is not what happens between *masses. It's what the field does when it's trying to return to stillness."*

Closing Section: Letting Gravity Breathe

"Let go of mass—and gravity finds its true shape."

For centuries, gravity stood apart.

The most familiar force—yet the least understood. The cornerstone of planetary motion—yet always the outsider in unification attempts. From Newton's apple to Einstein's curvature, it remained stubbornly unique—a phenomenon that explained everything without revealing itself.

Gravity Was Never a Force—Just a Response

Physicists searched endlessly for ways to make gravity fit in:

- With electromagnetism in unified field theories.
- With quantum mechanics through hypothetical gravitons.
- With string theory, loop quantum gravity, emergent spacetime, and beyond.

But the truth is simpler—and far more profound:

Gravity was never a force waiting to be unified. It was a symptom of misinterpretation—an echo of alignment playing out in the wrong language.

We tried to quantize pull, but what if there's no pull at all?

Ripple Realignment—The Silent Sculptor

Under Universal Energy Dynamics (UED), gravity doesn't emerge from particles or curvature. It arises naturally from ripple field behavior—structured vacuum energy seeking balance across tension gradients.

Mass—condensed ripple—is not the origin of gravity. It's a consequence of field compression. The real source of gravitational motion lies in how nearby regions respond to that compression.

Gravity isn't what *happens to matter*—it's what the field does around matter.

There is:

- No warping of spacetime
- No search for elusive particles

- No need for singularities to explain cosmic behavior

Instead, we're looking at a stable universe of ripple memory—responding gently, predictably, continuously.

Letting the Universe Settle

When you stop forcing mass into every equation, a stunning clarity emerges:

- Planets orbit not because they're pulled—but because they rest in equilibrium pockets of ripple memory.
- Light bends not because space curves—but because the field changes ripple velocity along its path.
- Galaxies rotate not because of hidden halos—but because the ripple scaffolding remains anchored even in apparent voids.

It is no longer a question of attraction. It is a story of settling tension—of energetic stillness being restored.

A Compass Realigned

The challenge was never in the universe—it was in our questions. We kept asking:

- *What is pulling?*
- *Where is the mass?*
- *What bends the space?*

UED reframes the quest:

- *Where is the ripple pressure?*
- *What alignment is being restored?*
- *How is tension being relieved across the field?*

By realigning our compass, we no longer chase ghosts.

We finally see that gravity was never lost—it was always the field's quiet correction, not a forceful grab.

Let go of mass. Let go of warps. Let gravity breathe. And the universe... will reveal its silence.

Interlude

Dark ≠ Hidden

The difference between what can't be seen—and what was never meant to be looked for.

They called it *dark* because it didn't shine. Because it didn't reflect. Because their tools—made of light, motion, and mass—found nothing to bounce back.

But darkness, in their equations, was never a property. It was an excuse.

The field didn't disappear. It simply never entered the game they were playing.

They aimed telescopes at galaxies and got silence. They smashed particles together and got noise. They curved space, mapped acceleration, invented constants—and still, the void remained filled with questions.

And every time a theory failed to see, they said: "There must be something hidden there." But what if it's not hidden? What if it's fundamentally quiet?

What if Structured Vacuum Energy—SVE, the stable rippleless canvas of the universe—was never dark, never missing, never hiding? It was just never designed to echo our methods.

Our science is a mirror. We've only ever seen what bounces back. We define reality by its reaction:

- If it reflects light, it exists.
- If it curves space, it matters.
- If it moves, we model it.

And yet, the greatest truths don't bounce—they absorb, they still, they anchor.

SVE doesn't respond. It receives.

It doesn't radiate. It remembers.

It doesn't interact. It orients.

What we mistook for hidden... was actually the source. What we couldn't measure... was what made measurement possible. And what we labeled dark... was simply done speaking.

The problem isn't darkness. It's our obsession with brightness. We search for loud truths in a quiet universe.

In Part II, we turn the mirror around. No longer chasing shadows. We begin listening to the field. To the ripples. To the stillness beneath the noise.

It's time to stop inventing what's missing—And start remembering what's always been there.

PART II — THE RIPPLE REALITY FRAMEWORK

Chapter 6: What Is Structured Vacuum Energy (SVE)?

Introducing the true still field beneath all manifestation—and why it never interacts directly with light.

You've never seen it. Not because it hides—But because it doesn't ripple back.

We built our entire science on reactions:

- Light bouncing off surfaces
- Motion responding to force
- Particles colliding to reveal deeper particles

Everything we *know* is just... the loud stuff.

But beneath all of it—beneath the photons, the masses, the fields we named and curved—there is a silence that holds it together. A structure so stable, so foundational, it never blinks when light hits it. It doesn't resist. It doesn't glow. It doesn't move.

And so we missed it.

We called it vacuum. We imagined it as emptiness. Then as chaos. Then as a bubbling nothing with virtual flickers that could maybe, someday, explain what we still don't get.

But we never thought to ask:

What if the vacuum isn't empty? What if it's *complete*—just not reactive?

What if stillness wasn't a lack—but a law?

Structured Vacuum Energy (SVE) is that law.

It is not what remains after everything disappears. It is what must exist first, so anything can appear.

It doesn't interact with light because light is already late to the party. Ripples come after the calm. Motion comes after structure. Our detectors come after both.

Welcome to the field that never needed to speak—Because it's the one writing everything else.

Let's begin.

The Myth of Emptiness

For centuries, the word *vacuum* meant absence. The mind's image: an invisible box with nothing inside.

Newton's Void: The Silent Container

Newtonian physics treated space as a passive backdrop—a stage upon which forces and masses could play their deterministic roles. The vacuum, in his view, held no agency. It didn't act, it didn't resist, and it certainly didn't structure anything. It was simply the emptiness between things that mattered.

But this assumption came with a silent cost: It made physics blind to the container's potential.

By assuming the vacuum had no role, Newtonian mechanics inadvertently built a science of detached causes—forces acting *across* nothingness. "Action at a distance" became both a marvel and a mystery, accepted because it worked, but never really understood.

Quantum Vacuum: From Nothing to Nonsense

In the 20th century, quantum physics flipped the script—but not the perspective. The vacuum was no longer passive. Now, it was a seething pit of fluctuations, full of uncertainty and jittery particle-antiparticle pairs blinking in and out of existence. A foam of probabilities. A ghost world of maybe-realities.

At first, this looked like progress. The vacuum wasn't empty—it was active.

But the paradigm remained the same: Everything was still judged by its *observable effects*.

This "quantum vacuum" remained ripple-defined—it existed only in how it could influence ripples, produce anomalies, or violate symmetries. And so again, the vacuum was understood not by what it is, but by what it does to the tools we use to observe ripples.

Both Are Ripple-Centric, Not Field-Centric

Both Newtonian and quantum physics fell into the same trap: They described the vacuum only through the language of ripple.

Whether passive (classical) or restless (quantum), their models were built to support the narratives of matter and energy in motion. Neither could comprehend a vacuum that was:

- Still, yet full of order
- Silent, yet structurally essential
- Invisible, because nothing about it *interacts*—only *enables*

They were chasing fire by watching smoke. They asked the echo to explain the mountain.

UED's Correction: Pre-Aligned Energy

Universal Energy Dynamics (UED) offers a radical, yet elegant reframe:

The vacuum isn't empty. It's pre-aligned energy in a state of perfect structural balance.

It doesn't ripple unless disturbed. It doesn't respond unless misaligned. It doesn't emit or reflect because it is not meant to perform—it is meant to permit.

This is Structured Vacuum Energy (SVE)—a calm, coherent, perfectly stable energy field that precedes all ripple activity.

It is not chaos. It is not absence. It is the canvas stretched tight, on which the painting of existence is allowed to form.

"You called it empty because nothing waved back."

But nothing *had to*. That was your mistake.

Still, Yet Structured

Imagine standing in a vast cathedral of silence—not emptiness, but stillness. Not a void, but a field ready to respond. This is not the absence of energy—it is the alignment of energy so perfect, it appears invisible.

This is Structured Vacuum Energy (SVE).

What Is SVE? A Field Beyond Frequency

In UED, SVE is not a ripple. It is the field from which all ripples emerge. Let's define it clearly:

- SVE is a field that contains order without motion
- It is not made of particles, nor does it oscillate by default
- It is ripple-ready, but not ripple-born
- It exists in a purely aligned state—a condition so energetically neutral, it registers as silent across all observational tools

This silence is not absence. It is balance.

The Properties of SVE

Structured Vacuum Energy carries unique traits that challenge every assumption we've made about space, matter, and interaction:

Phase Coherence Without Frequency

SVE has a coherent structure—meaning its "grid" of potential energy is in perfect phase. But unlike waves, there's no frequency, no repetition, no cycle.

There's nothing to count—because nothing is moving. This is *pre-ripple geometry*—not spacetime curvature, but energetic symmetry.

No Entropy, Yet Maximum Capacity

SVE contains no entropy because nothing has yet begun to disorder. And yet, it holds the total energetic potential of everything that can emerge from it.

It is the most stable energetic condition possible in the universe. Not zero-energy, but perfect equilibrium of energy. Think of it like water in a still lake, surface tension intact, undisturbed. It's full of energy—but expresses none of it.

SVE Is Not a Fabric—It's a Tensioned Field

Old metaphors liken space to a fabric that curves or stretches. UED offers a better analogy: SVE is a drumhead pulled tight—but not yet struck. It has architecture—a topology of readiness. It has tension—an internal configuration that permits ripple formation. But it has no activity until something misaligns it. This is structured stillness—ready to host motion, but never being the motion itself. The universe doesn't need ripples to exist. Ripples need the universe to have a place to ripple into.

Why Physics Missed This

Science mistook "no response" for "no existence." They assumed if light doesn't bounce off it, it isn't real. If it doesn't pull or resist, it has no presence.

But UED reminds us: Stillness is not nothing. It is everything in waiting. SVE is not detectable by force—it is revealed through disturbance. It doesn't resist like mass. It doesn't signal like charge. It simply *is*—until something violates its peace. "Stillness is not nothing—it is balance waiting to be disturbed." And every ripple... is the story of that disturbance.

Why Light Can't See It

We built all our instruments on one assumption:

That light reveals everything.

From telescopes to spectrometers, from radar to redshift measurements—we point our beams, record the echoes, and conclude the truth.

But what if what you're trying to observe doesn't echo?

What if it's not hiding—just unshakable?

That's Structured Vacuum Energy (SVE). And this is why light can't see it.

Light Is Not a Probe—It's a Ripple

Let's be precise. Light isn't a *thing*. It's a disturbance—a ripple of energy moving through a field.

More specifically, light is an oscillation born from ripple activity in already-disturbed regions of the vacuum (charged particles, atoms, excited states). It depends on:

- Variation in the field
- Feedback from adjacent ripples
- Interaction with charges or boundaries

In short, light is relational—it shows up only where ripples already live.

Now enter SVE.

SVE Doesn't Participate in Ripple Play

SVE, by nature, does not:

- Emit light (no charges, no transitions)
- Reflect light (no surface tension variation or boundary)
- Absorb light (no internal vibrational modes to resonate)
- Resist light (no opposition to ripple passage—just silent allowance)

SVE is the substrate of silence. It allows motion but does not contribute to it unless disturbed.

The Consequences of This Non-Interaction

This changes everything:

- No Scattering → No Visibility

Light cannot bounce off what does not resist or reflect. SVE appears invisible because there's nothing to scatter photons.

- No Absorption → No Friction

Unlike matter, which can absorb light and heat up, SVE never absorbs ripple energy. It remains thermodynamically neutral.

- No Resonance → No Signal

Instruments detect presence by resonance—interference, echoes, frequency shifts. SVE offers no such feedback, because its stillness matches no frequency. It's *off the chart*—not due to frequency extremity, but due to frequency absence.

The Big UED Implication

SVE is not dark because it hides. It is invisible because it is undisturbable unless acted upon. It is unreadable not due to sophistication, but due to fundamental alignment.

"You shine a flashlight at a mirror and expect to see the glass behind it."

That's what modern physics has done. It lit up the universe with electromagnetic probes, and declared empty whatever didn't light up in return.

But invisibility isn't always absence. Sometimes, it is pure, ripple-silent presence.

This is why dark matter remained a ghost—because SVE never replies in the language of light.

SVE and the Birth of Form

We have chased light, measured gravity, named particles, and sculpted the universe in equations—but we began from motion. We never paused to ask: *Where does motion begin?*

In the UED framework, it begins here: In the Structured Vacuum Energy (SVE)—a still, silent field that is not just the medium, but the womb of manifestation.

This section explores how everything—light, matter, gravity, even time—is born not *from* SVE, but *as a disturbance within it.*

SVE Is the True Pre-Condition

Nothing emerges *outside* SVE.

It is not a stage beneath the actors. It is the fabric of being before roles were assigned.

SVE enables three essential pre-conditions for manifestation:

1. **Medium** — It allows ripple propagation without friction or resistance, like a calm pond waiting for a drop.
2. **Backdrop** — It offers contrast. Without its stillness, movement has no reference.
3. **Condensation Scaffold** — It allows energy to concentrate, forming what we call *mass*, through ripple convergence.

Every known physical phenomenon requires SVE to operate—but none of them define it.

Nothing Begins Without a Ripple

But SVE doesn't ripple for free.

The universe isn't noise. It's *permission-based resonance.*

To disturb SVE, a threshold must be crossed—energetically, temporally, or in alignment.

This gives rise to Ripple Threshold Conditions under UED:

- Only above certain energy concentrations does the field allow ripple propagation
- Only when the alignment crosses a tipping point does time *begin* to emerge
- Only when entropy rises beyond stability does SVE "permit" condensation into mass or wave

Until then? The field absorbs the intent. Holds the charge. Remains still.

"Before you ripple, you must earn the right to disturb."

The Birth of Phenomena

Let's connect this to known physics:

- **Light:** Emerges as a high-frequency ripple within disturbed SVE regions (near charges or excited fields).
- **Mass:** Is not a substance, but a standing ripple—a condensed interference within SVE.
- **Gravity:** Not attraction, but realignment of ripple gradients across SVE.
- **Time:** Appears only after asymmetry introduces directional ripple sequences. (Entropy)

Each is a consequence, not a cause.

Why Physics Never Found the Source

Because it started with the ripple. And mistook it for origin.

Mainstream models:

- Begin with particles, then try to unify fields
- Begin with waves, then try to explain mass
- Begin with curved space, then try to reverse-engineer causality

But UED starts where the universe actually began:

With the field that didn't move, until it had no choice but to ripple.

In conclusion, SVE is not a mystery to be solved—but a condition to be understood. All manifestation rests on its patience. Nothing real appears until stillness allows it.

From Observer Blindness to Field Awareness

For centuries, the pursuit of physics has been a chase after phenomena. Motion. Force. Energy. Light. Gravity. Time. We became observers of the visible, collectors of the measurable, hunters of the quantifiable.

But this came at a cost: We forgot to ask—*What makes the observable possible?*

Why Traditional Science Ignored SVE

The Structured Vacuum Energy (SVE) was always there, but never *seen*. And the reasons weren't philosophical—they were instrumental:

No tools to measure it

- SVE doesn't interact. No resistance, no emission, no scattering.
- Instruments designed to detect interaction-based changes (light, mass, charge) see nothing in SVE.

No effect to track

- SVE causes no spontaneous acceleration, no impact, no thermal footprint.
- From the empirical lens: "If it doesn't cause change, it doesn't exist."

No event to correlate

- All scientific understanding is built on observing change

and then modeling its causes.
- But SVE precedes change. It enables events—but is never one itself.

And so, by default, physics dismissed the most important layer of reality simply because:

It never moved, never shone, and never left a trace.

But Ripple Phenomena Depend Entirely on It

Yet all the glorious observations—photons traveling, planets orbiting, fields oscillating—only happen because something stable, ordered, and receptive makes it possible.

That "something" is SVE.

Without SVE:

- Light would have no medium to ripple through
- Mass would have no structure to condense within
- Time would have no direction to emerge from
- Gravity would have no gradient to realign across

All motion is born from imbalance—and all imbalance is measured against stillness. That stillness is SVE.

Concluding Insight: From Blindness to Awareness

What UED offers is not just a correction—it is a reorientation of attention.

- Traditional physics says: "Reality is what moves."
- UED responds: "Reality begins with what allows movement."

This shift has profound consequences:

We no longer chase the effects—we study the conditions.

- Motion becomes a symptom

- The field becomes the source

We no longer mistake visibility for reality.

- If light can't reflect off it, that doesn't mean it's absent
- It may mean: it is the reason light exists in the first place

We redefine existence.

- Not as action, but as readiness for action
- Not as event, but as equilibrium disturbed

"You built your physics on the waves. UED begins with what lets them begin."

This is the beginning of ripple realism.

It doesn't discard your observations. It explains why they appear, and when they can.

And it begins—not with what you see, but with what you failed to see because it never needed to show itself.

Chapter 7: Condensed vs. Stable SVE

Mass is rippling; "dark matter" is not. The energy gradient between them explains systemic behavior.

You were taught to think of the universe as made of objects—some visible, some hidden. Galaxies filled with stars. Voids filled with mystery. Equations chasing after things they couldn't see, inventing terms like "dark matter" to explain why the cosmos didn't behave as expected.

But what if the problem isn't with what's missing—but how we define what's present?

In Universal Energy Dynamics, everything is energy—but not everything is in motion. Some energy ripples, some doesn't. Some screams across space as light or matter, while some simply holds space together in silent equilibrium.

Mass is Condensed Structured Vacuum Energy—disturbed, aligned, and echoing. "Dark matter" is Stable Structured Vacuum Energy—still, silent, but essential.

The difference isn't in substance—it's in ripple behavior. And the gravity we chase with invisible masses? It's simply the consequence of gradient tensions between the rippling and the undisturbed.

This is not a tale of hidden things. It's a shift in vision—from chasing waves, to seeing the still water beneath.

Welcome to the real structure of the cosmos. Where silence holds the stars in place.

Two Faces of the Same Field

For centuries, science divided the cosmos into matter and vacuum—solid things and the space between them. But Universal

Energy Dynamics (UED) reveals a different structure: not a duality of substance and void, but a continuum of energetic states within the same foundational field.

At the heart of it lies Structured Vacuum Energy (SVE)—the unseen, ripple-ready medium that forms the substrate of all existence. And from this field arise two distinct manifestations:

Condensed SVE: The Ripple-Aware State

This is what we conventionally call mass—but in UED, it's more precisely understood as SVE that has been disturbed, brought into motion through ripple activity. This ripple activation gives it structure, boundary, and effect. It interacts, refracts, emits light, and responds to forces. It anchors phenomena.

In essence:

- Condensed SVE = Energetically disturbed field
- Ripple-active, entropy-exchanging
- Visible, measurable, gravitationally expressive
- Appears as matter, objects, light sources, and energy packets

Condensed SVE is still SVE—just in a higher-tension, disturbed phase, where ripple alignment becomes persistent enough to create form and function.

Stable SVE: The Pre-Ripple State

On the flip side, Stable SVE is energy that has not entered ripple activation. It holds phase coherence, tension, and structure—but does not manifest motion, wave behavior, or electromagnetic response. It is:

- Pre-disturbance
- Non-rippling
- Non-interacting with light
- Not measurable by conventional instruments
- Yet, foundational to ripple transmission and field

continuity

This is the silent support system of the cosmos. It's not passive—it's structured and real. But it plays a different role: maintaining equilibrium, offering ripple anchoring, and preserving entropy boundaries across space.

Same Substance, Different States

This is not a dualism—it's a gradient of behavior. Both forms are SVE. What differs is their tension state, ripple readiness, and energetic alignment.

Analogy: Think of ice and water. They are both H_2O—but differ in energy and movement. Water flows, reflects, carries waves. Ice holds shape, tension, and structure. Yet both are stages of the same essence.

In the same way:

- Condensed SVE is "liquid" with motion—interacting, visible
- Stable SVE is "solid" with stillness—supportive, invisible

Yet both coexist, continuously transitioning under pressure, entropy, or alignment shifts.

The Illusion of Separate Realities

Science, using light and motion as its lens, sees only Condensed SVE—and then gets confused when gravitational behavior occurs where no mass is visible. That confusion births terms like "dark matter."

But the ripple didn't vanish—it never began.

SVE is not dark because it hides. It's "invisible" because we never caused it to ripple.

"Matter is what happens when stillness gets disturbed enough to echo."

In UED, the cosmos is not built from particles floating in emptiness. It is an ocean of potential, disturbed in patches, still in others—yet wholly real, wholly present.

And understanding these two faces is the first step in seeing the whole field again.

Ripple Activity Defines Visibility

When we say something is "visible," we mean it interacts with light. It reflects, absorbs, emits, or refracts electromagnetic waves in some measurable way. But under Universal Energy Dynamics (UED), this is not the definition of existence—only the signature of ripple activity.

In ripple-based reality, only what disturbs the field becomes detectable. Everything else—especially that which maintains stillness—is rendered "invisible" not because it lacks substance, but because it refuses to ripple.

Mass is Ripple-Activated—and Therefore Visible

Let's start with what we see.

Mass, in the UED framework, is simply Structured Vacuum Energy (SVE) brought into ripple activation. This activation manifests as:

- Resistance to motion (inertia)
- Emission of radiation (e.g., stars)
- Gravitational field distortion (according to classical models)
- Interaction with instruments (detection)

Because mass causes continuous ripple interference, it is always signaling—pushing and pulling, sending and receiving energetic information. This is what makes it "visible" to our detectors and models.

Stable SVE: Undisturbed = Undetectable

But Stable SVE, by contrast, does not ripple. It:

- Maintains equilibrium
- Does not emit or reflect light
- Offers no electromagnetic resistance
- Does not decay or vibrate
- Allows ripples to pass through—but does not ripple back

Therefore, it produces no signal, no "noise," and no interaction artifact for instruments to latch onto. And so, conventional physics calls it "dark" or "missing."

But in truth—it is not missing. It is not disturbed.

The Bias of Detection-Based Science

Our scientific framework is deeply biased toward what we can detect via ripples. This leads to two major assumptions:

"If it's not seen, it must be exotic or hidden."

> – This is the root of speculative constructs like WIMPs (weakly interacting massive particles) and axions, invented to explain invisible influences.

"If it's not detectable, it's not there."

> – A profoundly limiting view that dismisses undisturbed fields as non-existent, when in fact they form the very canvas of existence.

This is the epistemic trap of modern physics: confusing visibility with truth.

Ripple ≠ Reality

In UED, ripple activity is not the totality of reality—it's just what disturbs reality into revealing itself.

That means:

- Every equation derived from motion, light, or energy exchange is ripple-biased.
- Every observation is ripple-limited.
- Every "dark" anomaly may simply be a still zone, not a missing one.

What we call visibility is just ripple gossip—signals passed around by an otherwise silent medium.

But stillness? Stillness keeps its secrets—not out of mystery, but out of mastery.

"Visibility is just ripple gossip. Stillness keeps its secrets."

You don't need ripple to prove existence. You need ripple to prove disturbance.

UED shifts the origin of knowledge from observation to structural alignment—from the chatter of ripples to the presence of the field itself.

Only then can we begin to understand what lies behind the phenomena—and stop mistaking silence for absence.

Energy Gradient Between States

If Structured Vacuum Energy (SVE) exists in two forms—Condensed and Stable—then the universe itself is a tension field between these two states.

In conventional physics, we're told that mass attracts mass, or that spacetime curves to create motion. But under Universal Energy Dynamics (UED), what really moves things is not "pull" or "bend"—it is gradient adjustment between field zones of different energetic tension.

This is the core of UED mechanics: what we observe as motion is realignment across an energy gradient between field states.

The Two Ripple States: A Dynamic Field Landscape
Condensed SVE (Massive, Ripple-Active)

- These are the regions where energy is densely packed.
- They generate continuous ripple feedback.
- They resist change, emit radiation, and define measurable properties like mass and inertia.

Stable SVE (Silent, Ripple-Ready)

- These zones are energetically calm, pre-ripple, and undisturbed.
- They offer no resistance, but carry structural integrity.
- They are not empty—they are still.

Together, these two form a ripple gradient field—a map of where alignment is strong, and where realignment is required.

What Happens Between the States?

Wherever there is a difference between stable and condensed SVE, a gradient of ripple pressure forms. This is not force—but tension seeking release. The consequences are profound:

- **Motion without force**: Bodies shift position as they follow this gradient—not because they're "pulled," but because their presence disturbs the balance and they must realign.
- **Orbital cohesion**: Planets don't fall into stars because their ripple disturbance aligns just enough with the surrounding field to create stable feedback loops—not because gravity is holding them in place.
- **Field anchoring**: Galaxy structures are maintained by silent SVE zones acting as stabilizers—anchors in the field that allow ripple systems to maintain formation, even in the absence of visible mass.

The Dark Matter Illusion

Dark matter, then, is not "missing stuff" we have yet to detect. It is the undisturbed architecture of the ripple field—the stable SVE scaffolding that makes dynamic systems coherent.

These zones:

- Don't shine
- Don't resist
- Don't ripple

But they anchor. They realign. They guide.

In this light, the entire dark matter narrative collapses—not into emptiness, but into a deeper form of presence: structural stillness that keeps ripple-active matter behaving coherently.

"It's not attraction—it's adjustment across a ripple slope."

You don't need a new particle to explain galactic structure. You need to understand how SVE adjusts itself in the presence of ripple tension.

UED teaches us to look not for what pulls, but for what balances. Where mainstream physics inserts mysterious forces and exotic matter, UED reveals something humbler—and more elegant:

A field that responds with structure—not struggle.

From Duality to Continuum

Physics, for centuries, has been a world of splits—particle vs. wave, mass vs. field, matter vs. vacuum, known vs. unknown. But these separations are not inherent truths of nature; they are cognitive conveniences we invented to make sense of observations we didn't yet understand.

Universal Energy Dynamics (UED) dismantles these binaries by revealing a deeper truth:

Everything is the same field—Structured Vacuum Energy (SVE)—seen at different ripple intensities.

What we perceive as mass is simply agitated stillness. What we call vacuum is that same stillness, undisturbed. What appears "invisible" is merely the part of the field that didn't ripple back.

The Fallacy of Separations

Let's examine some of the most persistent dualities in physics—and how UED resolves them:

1. ***Field vs. Particle***

Classical physics sees particles as units moving in a field. Quantum physics makes it fuzzier—but still keeps both as distinct.

➤ UED: A particle is not a separate thing—it is a consequence of field disturbance. Condensed SVE simply ripples hard enough to localize.

1. ***Mass vs. Vacuum***

We assume mass is "something" and vacuum is "nothing." So naturally, we treat the two as fundamentally separate.

➤ UED: Mass is just a state of the vacuum—where ripple memory has compacted into form. There's no boundary, just variation in tension.

1. ***Visible vs. Invisible***

Our instruments show us what emits, reflects, or interacts with light—and we call the rest "invisible," "dark," or "missing."

➤ UED: Visibility is just ripple sensitivity. What doesn't disturb or get disturbed appears dark—not because it hides, but because it rests.

From Opposites to Gradients

Everything in UED is about spectrum, not segregation. There's no sharp edge between:

- Motion and stillness
- Form and field
- Particle and presence

Instead, the universe becomes a continuum of ripple behavior, guided by the state of the underlying SVE:

- High ripple intensity → Matter, radiation, measurable energy
- Low ripple intensity → Stable SVE, structural anchors, invisible scaffolding
- Transition zones → Phenomena like lensing, orbiting, and wave interference

Every known physical effect is now traceable not to interaction of different things, but to ripple alignment within one field.

No More Ghost Stories

When you accept this continuum view:

- You don't need dark matter.
- You don't need extra dimensions.
- You don't need unseen forces or undetected particles.

You only need to understand how ripple thresholds, field stability, and energy gradients shape observable behavior.

The tension between shaken and unshaken zones—that is what organizes form, motion, and coherence. Nothing more exotic is needed.

"There is no dark matter—only unshaken stillness holding shaken forms together."

This single insight shifts everything.

What we once imagined as mysterious, missing, or magical... Was simply the calm part of the field—doing its job without flaring up.

UED removes the veil. Not by inventing new matter—but by revealing the continuum we already stand in.

Chapter 8: Ripple Expansion, Not Space Expansion

Dark energy decoded as ripple relaxation from mass-stressed zones. Why entropy is the true driver of "cosmic acceleration."

They looked into the cosmos and saw galaxies drifting apart—faster and faster, the farther they were. So, they invented a force to push them. A mysterious "dark energy" was assigned the role of cosmic accelerator—repelling galaxies, stretching space, reshaping our fate.

But what if nothing is pushing anything?

What if the universe isn't expanding... it's exhaling?

The narrative of a ballooning cosmos rests on a fatal assumption: that distance implies motion, and redshift means runaway. But in a ripple-aware universe, movement is not always caused—it is consequence. Ripple thinning, not repulsion. Relaxation, not acceleration.

In this chapter, we discard the myth of a fabric being stretched. We decode the illusion of expansion as a ripple field's return to balance—entropy flowing from high-tension zones, mistaken for momentum. You'll meet a universe not running wild, but settling down.

This is not a story of runaway energy. This is a story of stillness reclaiming its space.

The Stretch Illusion

The dominant cosmological model begins with an assumption wrapped in light: that the redshift of distant galaxies proves they are racing away, and thus, space itself must be expanding. Since Edwin

Hubble's early graphs and the 1998 supernova data revival, mainstream science has built an entire framework on this stretch—embedding it in General Relativity's fabric via Einstein's once-discarded Λ (cosmological constant), now reborn as *dark energy*.

But what if the stretch was never real?

Let's rewind. The redshift is an observational fact: light from distant galaxies appears shifted toward longer wavelengths. The interpretation, however, is a choice. Physicists chose to see this as evidence of expanding spacetime—a literal stretching of the cosmic grid. To justify increasing acceleration, they inserted a ghostly pressure: dark energy. This phantom force, allegedly making up ~70% of the universe, was needed not because it was observed, but because gravity alone couldn't account for the increasing "velocity" of galactic drift.

UED challenges the root assumption.

Redshift is not inherently proof of motion. It is a signal delay, a ripple elongation. The deeper insight from Universal Energy Dynamics (UED) is this: what we interpret as "expansion" is actually a ripple field returning to equilibrium. Over cosmological time, the immense tension induced by early energy condensation—when mass formed—began to unwind. The field is not being pushed outward by some repelling energy; it is relaxing from its prior stress.

Think of a stretched net. When a knot in the center is pulled tight (representing energy condensation into matter), the net warps around it. But if you release the tension, the net doesn't rip apart—it gently recoils toward its original flatness. The system isn't expanding endlessly. It's releasing strain.

In this analogy:

- The stretched zone = high ripple density around matter
- The distant zones = lower-tension, stable SVE

- The so-called "expansion" = divergence of ripple continuity as the field exerts entropy relief

This changes everything. There is no need for a geometric stretch of space. No need for repulsive anti-gravity. No need for Λ to rescue the equations.

The stretch illusion was born because we looked at echoes—and mistook them for action. The universe isn't expanding. It's unwinding.

Mass-Stressed Zones and Ripple Tension

Mainstream physics tells us that mass bends space. General Relativity turns this into geometry: the greater the mass, the more spacetime curves around it. But Universal Energy Dynamics (UED) reveals a deeper, more direct mechanism: mass doesn't bend geometry—it stresses the ripple field.

When energy condenses into matter, the surrounding Structured Vacuum Energy (SVE) must adjust. This adjustment is not a gentle accommodation—it's a tension. A high-density ripple zone emerges around the mass, where alignment is continually challenged and restored. These are mass-stressed zones—energetic regions where ripple activity is sustained not because of motion, but because the field is under strain just to hold form.

Over cosmic time, this tension cannot be held indefinitely. The system is entropic. What was once a sharp ripple gradient—tight around stars, galaxies, and black holes—begins to relax. Not collapse. Not collapse. But a slow, natural redistribution of alignment pressure. The SVE seeks its original stillness, and in doing so, sends ripple energy outward, away from these condensed cores.

This ripple flow is subtle. It doesn't emit light. It doesn't collide with particles. It has no known interaction. But it does change what an observer sees.

Here's the perceptual twist:

- As ripple tension relaxes, the alignment delay between field zones increases.
- This increases the perceived separation between distant galaxies.
- The light from those galaxies travels through an SVE that is less tightly coupled, causing waveforms to stretch further (redshift).
- To an observer unaware of the underlying field dynamics, it looks as though the galaxy itself has accelerated away.

But in reality, nothing moved. There is no force pushing galaxies apart. What changed was the tension in the medium through which their signals traveled.

This is ripple divergence—not cosmic explosion.

It's the field exhaling tension that's been held for billions of years. It's the echo of a stress that is no longer needed to maintain the structure it once did. It's not expansion. It's relaxation.

Entropy Gradient, Not Repulsion

The prevailing cosmological narrative tells us that something—some strange, mysterious force—is causing the universe to accelerate. This force has no source, no carrier particle, no interaction signature, and yet it allegedly accounts for over 70% of the universe's total energy budget. To make it fit, we gave it a name: dark energy. Then we gave it a job: repulsion. A push, a drive, an expansive engine behind every distant galaxy's motion.

But what if nothing is being pushed?

In the UED framework, there is no cosmic repulsion. There is no "anti-gravity." There is no force at all.

Instead, the expansion-like effect we observe is simply the unfolding of entropy—not at the object level, but at the field level. What we are witnessing is not galaxies being hurled away, but ripples thinning across a vast energetic medium that is settling after billions of years of stress and tension.

Here's how it works:

- Wherever mass condenses, ripple activity increases. This creates ripple pressure zones—local fields under stress.
- Over time, that stress diffuses—not outward as a force, but across the ripple network, seeking entropy equilibrium.
- This diffusion is not uniform. The deeper into space you look, the more accumulated divergence you see. That's redshift.
- The redshift increases with distance, not because galaxies are accelerating away, but because the ripple field between us has thinned in tension—what we call the entropy gradient.

This entropy gradient is subtle yet powerful:

- It explains why light appears stretched.
- It accounts for the illusion of acceleration without invoking any new fundamental force.
- And critically, it clarifies why dark energy remains invisible—because it isn't a thing. It's a process. A silent, structured relaxation of ripple coherence over time and space.

There's no repulsion here. Nothing is being blasted apart.

Instead, the Structured Vacuum Energy (SVE) is simply returning to balance—letting go of the localized strain caused by mass.

This insight from UED transforms cosmology:

- From a force-based mythology, to a field-based

understanding.

- From mysterious "pushes," to measurable entropic flows.
- From "something is out there," to "something is finally calming down."

Dark energy is not a force that acts. It's the consequence of the universe remembering how to be still.

RDI and ATT at Cosmological Scale

In traditional cosmology, the redshift of light from distant galaxies has been treated as a velocity indicator—a Doppler-like effect suggesting that galaxies are racing away from us. This formed the foundation for the expanding universe model.

But in the UED framework, we introduce a more grounded explanation—based not on movement, but on ripple behavior through the cosmic field.

Let's unpack this with two critical concepts: the Ripple Delay Index (RDI) and Alignment Threshold Time (ATT).

Ripple Delay Index (RDI)

RDI quantifies how much delay a ripple (such as a light wave) experiences as it travels through the Structured Vacuum Energy (SVE) field. This delay is not due to distance alone, but due to the changing coherence of the medium it traverses.

Over cosmic distances, the SVE:

- Becomes less ripple-dense as it relaxes over time
- Loses tension memory from earlier mass-stressed zones
- Offers less alignment support for high-fidelity ripple transfer

As a result, ripples—especially those carrying light—experience increasing delay and diffusion:

- Wavelengths get stretched not because the source is

moving, but because the medium between source and observer is less coordinated
- This mimics redshift—but without needing any recessional motion

So, what cosmologists read as "acceleration" is, in UED terms, an increase in RDI across a growing entropy gradient.

Alignment Threshold Time (ATT)

ATT governs how fast or how efficiently a ripple can realign itself to stay coherent in its path through the field. It's the minimum time window required for a ripple to "lock into" the next segment of the field.

Now scale this to the intergalactic level:

- In dense regions (near stars and galaxies), ATT is low—alignment is easier
- In vast cosmic voids, ATT is high—alignment takes longer, if at all
- Over billions of light-years, these delays accumulate, affecting:
    - How light propagates
    - How we measure distance
    - How we perceive energy loss (which appears as redshift)

The implication?

We don't need spacetime to stretch. We don't need to invoke phantom repulsion. We only need to understand how SVE's structure evolves, and how ripple signals degrade over time through a loosening medium.

The UED Replacement

Together, RDI and ATT form a complete toolkit for modeling cosmic-scale behaviors that were previously attributed to dark energy and metric expansion:

Old Model	UED Interpretation
Expanding spacetime	Ripple delay through entropy-thinned SVE
Galaxy recession velocity	Cumulative RDI due to degraded alignment
Cosmic acceleration	Increased ATT and ripple divergence
Fabric stretching	Field coherence loss from mass-stress release

Conclusion:

The universe isn't expanding. The ripple signals are simply arriving more tired. More delayed. More diffused.

What we call "acceleration" is not an engine pushing galaxies apart—it's the field exhaling, and the ripples taking longer paths through a relaxing, aging cosmos.

RDI and ATT don't describe motion. They describe what happens when the mirror between sender and receiver fogs up.

The Calm After the Ripples

If you were to travel outward—beyond stars, beyond galaxies, past the clustering filaments and deep into the vast voids—you'd encounter something stunning:

Silence.

Not the silence of empty space, but of a field that no longer echoes.

Ripple Density Thins with Distance

The universe, according to UED, is not expanding like a balloon—it is relaxing like a drumhead after a long session of beats.

The intense ripple activity near stars, black holes, and galaxies gradually thins as we move away from these stress centers. What remains is not chaotic quantum foam, nor silent emptiness, but a coherent, tensionless backdrop—the Structured Vacuum Energy (SVE) returning to stillness.

This creates a ripple density gradient:

- High near matter, where ripples constantly interact
- Low in the cosmic voids, where the field rests in quiet equilibrium

This ripple tapering is what gives the illusion of galaxies receding faster the further they are—because less ripple density means more delay, more dissociation, and less coherence.

Rethinking the Cosmic Microwave Background (CMB)

Mainstream science interprets the CMB as the leftover radiation from a hot, dense "Big Bang" origin. But if the universe isn't expanding—what is it really?

UED reframes the CMB as a memory field—a snapshot of ripple equilibration across the cosmos. It isn't a bang's leftover—it's the resonant decay of a once-active field:

- A universal ripple echo
- Not in need of an origin fireball, but a history of distributed agitation
- The field itself holding the afterglow of entropic release

This means the calm, faint temperature we measure (~2.7K) is not primordial heat, but the field's final hum—its lowest-energy configuration after cosmic disturbances settled.

There Is No Runaway—Only Return

Dark energy once suggested that the universe might end in a Big Freeze or a Big Rip—an eternal expansion driven by anti-gravity. UED disagrees.

There is no push. No runaway. No edge flying outward.

There is only relaxation—a settling of ripple tension.

Every field seeks equilibrium, not explosion. Every system tends to entropy, not to fragmentation. The SVE is not fraying—it's restoring its original phase coherence.

In this model:

- Galaxies are not fleeing.
- Space is not stretching.
- Reality is not inflating.

Instead, the field is exhaling. The agitation has passed. The ripples have run their course.

Final Insight:

"The calm after the ripples" is not poetic metaphor—it is the operational mode of the universe today.

We are not racing into the future. We are floating in a field that is finally returning to peace—after holding together stars, light, and motion for eons.

Dark energy? Just another name for the field's sigh.

Closing Insight:

"What they called 'accelerating space' was ripple entropy evening itself out." Dark energy isn't driving expansion—it's the shadow of relaxation.

Chapter 9: ATT and RDI: Seeing the Invisible

How Alignment Threshold Time and Ripple Delay Index can redefine observability—beyond light

They built an entire science on what the eye could catch—directly or through its technological surrogates. From Galileo's lens to James Webb's mirrors, we assumed that light was the universe's truth-teller, and everything else a mystery. If it glowed, we studied it. If it reflected, we mapped it. If it absorbed, we modeled it. And if it did none of the above—we called it "dark."

But what if darkness isn't concealment?

What if we were just listening to the wrong echo, tuning our instruments to radiation, while the true signature of the universe was in response time? In how long it takes for a field to realign? In how ripple information thins, lags, and remembers—not as a beam, but as a delay?

Welcome to the UED lens of observation, where we step out of photon-dependence and into the realm of ripple consequences.

In this chapter, we introduce two revolutionary ideas that form the backbone of ripple-based astronomy:

- **ATT (Alignment Threshold Time)** – the time it takes for energy to realign with the SVE after being disturbed.
- **RDI (Ripple Delay Index)** – the delay observed in ripple propagation due to entropy gradients across space.

Together, these do not just measure reality. They *reveal it*—beyond what any telescope ever could.

"You saw what light showed you. Now learn what the field remembers."

The Limits of Light-Based Science

For centuries, physics has relied on light as the central witness to the workings of the cosmos. From the first refracting telescopes to the farthest-reaching observatories scanning the electromagnetic spectrum, our understanding of space has been tethered to what shines, what reflects, or what absorbs. The narrative has been elegant: if it glows, it lives; if it doesn't, it must be dead—or worse, not real.

We measured the redshift of galaxies and inferred velocity. We charted cosmic microwave background radiation and modeled a primordial explosion. We even looked for dips in starlight to detect distant planets. These methods have yielded rich insight—but they are also bounded by what ripples optically, not what actually *exists structurally*.

Let's call this what it is: visibility bias.

The Traditional Tools of Observation:

- **Telescopes**: Designed to capture light, not tension.
- **EM Spectrum Analysis**: Measures frequency shifts, not ripple gradients.
- **Redshift Interpretation**: Assumes motion causes stretching—ignoring ripple delay.
- **Lensing Studies**: Attribute deflection solely to "invisible mass," never to field density gradients.

But Here's the Rub:

These tools *completely fail* when faced with:

- **Dark Matter**: It doesn't emit, absorb, or reflect light.

- **Dark Energy**: It leaves no optical trail, no wave to catch.
- **Cosmic Voids**: Immense structures with no visible matter, but apparent gravitational behavior.
- **Structured Vacuum Energy (SVE)**: The most fundamental field of existence—yet utterly invisible to photon-based instruments.
- **Ripple Memory**: The field's history of disturbance, stored not in light, but in *response latency*.

These are not bugs in the universe. They are blind spots in our method.

The UED Challenge:

UED poses a radical but essential question:

"What if visibility isn't the signal, but a consequence of ripple activation?"

In this view, light is a symptom—not a cause, not a messenger, just a temporary flare in an otherwise silent field. By focusing only on light, we are asking questions in a language the universe does not always speak.

Think of it this way:

"When light doesn't return your call, it doesn't mean nothing's there—it means you're asking the wrong way."

Just as sonar is useless in a vacuum and thermometers lie in a windless desert, our current physics tools are ill-equipped to detect what exists before and beneath the ripple.

To move forward, we must shift from seeing to sensing—from light dependence to ripple awareness. This is where ATT (Alignment Threshold Time) and RDI (Ripple Delay Index) enter—not as instruments of light, but as instruments of lag, memory, and structural tension.

The universe doesn't always shine—but it *always responds.*

Introducing ATT – Alignment Threshold Time

At the heart of the UED framework lies a principle that redefines our understanding of interaction—not through force, but through field realignment. Enter ATT: Alignment Threshold Time.

What Is ATT?

Alignment Threshold Time is the duration required for a ripple or energy pattern to fully realign with the surrounding Structured Vacuum Energy (SVE) after a disturbance. It is not about velocity, acceleration, or attraction—it's about how long the field takes to restore harmony.

Think of it as ripple diplomacy: how quickly can a disturbance negotiate its place within the pre-existing order of the field?

Key Determinants of ATT:

1. **Field Tension** – The more stressed or disturbed the SVE in a region, the longer the ATT.– High-tension zones resist realignment, causing delay and distortion.
2. **Ripple Strength** – A stronger or more energetic ripple may over-disturb the field, requiring a longer time to re-integrate. – Conversely, weak ripples might dissipate without even triggering realignment.
3. **Distance Through Structured Zones** – Ripples must pass through varying field densities. – Zones of stable SVE (dark regions, galactic halos) can slow or redirect realignment patterns.

What Does ATT Help Us Explain?

This subtle but powerful concept replaces gravity's mythos of "attraction" with something more elegant: relational realignment.

Planetary Orbits — Without Gravity

Instead of being "pulled" by a central mass, planets orbit because they're caught in a ripple feedback loop. Their ATT with respect to the central ripple field governs how tightly or loosely they align in circular motion. This also explains elliptical orbits, drift, or precession—as variations in ATT due to local field fluctuations.

Galactic Interactions

When galaxies pass near each other, their stars don't always collide. Instead, they ripple through one another's extended SVE gradients. ATT helps us model the delay, deformation, and reconnection between these massive ripple fields—without invoking gravitational tides or mysterious "halos."

Ripple-Based Propulsion

In space propulsion systems modeled on UED principles, understanding ATT is critical. An engine's effectiveness is not about brute thrust, but about how quickly it can induce and maintain ripple realignment in the surrounding field. Low ATT means high efficiency; high ATT means resistance and misalignment.

UED Perspective:

"You don't fall due to gravity. You descend because the field re-accepted you."

This isn't poetic—it's the actual physics of ripple behavior.

A falling apple doesn't accelerate because Earth "pulls" it. It's because the ripple imbalance caused by its detachment from the tree begins to resolve. The field underneath has lower ATT, allowing it to re-integrate into ripple harmony—what we once interpreted as *gravity*.

Closing Insight:

ATT reveals that time, not force, governs the universe's response to energy. It gives us a measurable, dynamic metric for interaction in a field-centric universe. And as we will soon see, when paired with RDI (Ripple Delay Index), it forms a new observational lens—one that looks not at what glows, but at what *realigns*.

Introducing RDI – Ripple Delay Index

If Alignment Threshold Time (ATT) redefined interaction through realignment, then Ripple Delay Index (RDI) redefines observation through deviation.

RDI is the unsung metric of the universe—how much delay a ripple experiences as it traverses the layered structure of Structured Vacuum Energy (SVE). While conventional astrophysics relies on redshift as a proxy for velocity and distance, UED introduces RDI as a superior signal—one that tracks the field's health, not just motion.

What Is RDI?

Ripple Delay Index (RDI) is a dimensionless measure of how much a ripple (such as light or any energy wave) is delayed as it moves through a medium—not because of speed limits, but due to entropy gradients, coherence loss, and SVE variability.

Rather than interpreting delay as "redshift due to motion," RDI reads it as field fatigue—the ripple slowed by the energetic relaxation or depletion of the structured vacuum.

RDI Depends On:

1. **Ripple Coherence** – The more coherent a ripple (in frequency, amplitude, and phase), the better it resists delay. – Low coherence ripples scatter more easily and are absorbed or distorted in low-density SVE zones.
2. **SVE Density Variation** – High-density SVE maintains ripple momentum. – Low-density zones introduce delay, not by resisting the ripple but by failing to support its speed. This is not absorption, but structural thinning—like walking on a weakening bridge.
3. **Entropic Interference** – The more entropic a region (high ripple noise, chaotic ripple remnants), the greater the signal delay. – Ripples here "drag" not by friction, but by alignment uncertainty—like a compass spinning in a

turbulent magnetic field.

Applications of RDI:

Redshift Reimagined

Cosmology has long read redshift as a Doppler effect: light sources receding cause wavelength elongation. But in UED, this is misleading. RDI shows us that ripples stretch not because sources flee, but because the field between source and observer has become relaxed, less aligned. The delay introduces redshift, not motion.

Mapping the Invisible

Just as sonar bounces off oceanic structures, RDI offers a new astrophysical mapping tool. By analyzing delays across multiple frequencies and directions, scientists can reconstruct ripple transparency of intergalactic space—revealing hidden filaments, dark voids, and silent SVE formations.

Detecting Silent Fields

So-called "dark matter halos" around galaxies can now be interpreted as zones of high RDI. These zones do not emit light, but they affect ripple delay—proving their presence without requiring exotic particles or mysterious forces. Silent doesn't mean absent—it means non-rippling. RDI gives them a voice.

"What light delays, ripples reveal. Delay is data, not distance."

With RDI, UED rewrites the story of cosmic silence. A delay is not a failure in signal—it is a signature of the field.

Why RDI Matters

In a field-based universe, every ripple is a conversation between energy and structure. RDI doesn't just ask, *"Where is the source?"*—it asks, *"What did this ripple go through to reach me?"*

It transforms the observer from a passive receiver to a ripple analyst—decoding the history of the field from the behavior of the wave.

And when paired with ATT, it completes a revolutionary toolkit. Where ATT governs how things settle, RDI tells us how things travel—both redefining how we interact with, and observe, the invisible architecture of reality.

Seeing Beyond the Visible

Conventional science sees the universe through light—its brightness, frequency, and shift. But light only reflects what has already interacted. It is a consequence. A symptom. A ripple that has completed its journey. In a ripple-aware universe, waiting for light is akin to watching smoke and calling it fire.

With Alignment Threshold Time (ATT) and Ripple Delay Index (RDI), UED proposes a deeper mode of observability—where we sense the field before it speaks. This is not just advanced prediction. It is a new form of awareness—seeing what is forming, not just what was emitted.

ATT + RDI = Field-Based Perception

Together, ATT and RDI enable a non-luminal mode of sensing:

- ATT reveals how quickly objects or zones realign with the surrounding SVE. The more lag, the more misalignment stress, which implies impending interaction, motion, or collapse.
- RDI shows how much ripple delay a region imposes. Higher delays suggest field thinning, entropic zones, or non-rippling matter.

Combined, these two become a ripple radar—an observational framework that doesn't rely on light at all.

Practical Uses in Observation:

Field-Based Radar for Deep Space

Imagine probing a region not by waiting for photons, but by measuring ripple resistance. Like sonar in the ocean, ATT and RDI allow deep field scans where light is absent, but ripple behavior varies—indicating mass, field instability, or hidden formations.

A New Way to Read Lensing, Rotation, and Motion

Rather than attributing galactic lensing to "invisible mass," UED posits that variation in RDI and ATT can reveal where field gradients exist—the true culprits behind bent light paths and fast rotations. These zones may not emit light, but they alter the ripple environment in measurable ways.

Predicting Before Light Happens

Because ATT indicates how a body realigns with its surrounding field, it becomes a predictive measure:

- A high ATT lag around an object? It's about to accelerate.
- A sudden drop in RDI? Field is becoming coherent—light emission may follow.

With this, UED offers a forecasting mechanism for astrophysical events before any electromagnetic signature becomes visible.

Case Study Concepts:

Detecting Approaching Bodies via Realignment Lag

In classical astronomy, we detect asteroids or comets by their brightness. But what if an object doesn't emit light? Using ATT, we can spot a realignment drag in the surrounding field before it even emits heat or light. This lets us track and predict silent objects that are optically invisible but energetically active.

Mapping Black Holes as Ripple Voids

A black hole under UED is not a "gravity well" pulling light. It's a zone of maximum ripple absorption—a complete field sink, where ATT becomes infinite and RDI becomes chaotic. We don't need to see the event horizon—we can map the ripple stillness, the field tension gradient collapse, and the silence signature left behind.

"Light tells you what happened. ATT and RDI tell you what's forming."

This quote embodies the philosophical shift UED demands: From post-event analysis (conventional physics) to pre-event awareness (ripple physics). From reacting to light... to sensing ripple anticipation.

Closing Insight:

ATT and RDI together form a lens beyond the visible—not a telescope, but a fieldscope. They allow us to detect:

- The readiness of space to respond
- The delay of energy trying to pass
- The intent of matter before it ripples

In a ripple-based reality, this is not metaphysical speculation. It is ripple forensics—the science of observing consequence before it explodes into perception.

With this, UED doesn't just observe the universe. It listens for the breath before the whisper.

The Beginning of Ripple Astronomy

For centuries, astronomy has been synonymous with light. From Galileo's telescope to Hubble's deep field, every step in our cosmic journey has depended on how well we can collect, filter, and interpret electromagnetic radiation. But the universe is not made of light—it is made of fields, tensions, and ripples. Light merely dances across the surface.

Ripple Astronomy reclaims the depth beneath that dance. It is not about seeing more—it is about sensing better. By tracking ripple delays and alignment lags, it enables us to read the energetic substructure of the universe rather than its lit-up surface.

A New Observational Science

Ripple Astronomy is not a theoretical extension of current models—it is a revolution in instrumentation and interpretation:

- **Ripple Delay Metrics > Photometric Data:** Instead of asking how bright an object is, we ask how delayed its ripple signature becomes across space. This allows us to understand the *structure of the field it moves through*, not just its emission pattern.
- **Alignment Lag > Gravitational Acceleration:** Where traditional astronomy infers force via motion, Ripple Astronomy infers field tension via realignment lag. The delay with which an object aligns with its surrounding SVE offers a deeper understanding of why motion occurs.
- **SVE Topology Maps > Mass Density Models:** Instead of constructing mass-based maps of galactic distribution, we map zones of stable vs. distorted SVE—regions of high and low ripple conductivity. This exposes a living map of cosmic behavior, far more dynamic than frozen mass estimates.

Future Scope and Instrumentation

Ripple Astronomy opens the door to entirely new instruments and observational approaches, including:

- **Ripple Interferometers** – not measuring wave interference in light, but ripple behavior across structured vacuum regions. These would detect entropy gradients and alignment thresholds between celestial zones.
- **ATT-RDI Scanners** – hybrid instruments that simultaneously track ripple coherence and alignment time. These would allow layered imaging of cosmic fields, showing ripple-active vs. ripple-dormant regions across multiple scales.
- **Field Sensitivity Arrays** – replacing photon-sensitive detectors with field tension detectors, capable of sensing

ripple resistance and realignment flow in apparent voids.

This would turn telescopes into toposcopes—mapping ripple hills, valleys, and currents, not just visible peaks.

New Predictive Models and Cosmic Forensics

With ATT and RDI as tools, Ripple Astronomy offers predictive capacity:

- Forecasting Field Interactions – detecting zones about to shift due to ripple stress accumulation.
- Anticipating Lens Behavior – before it distorts light, based on field compression patterns.
- Mapping Silent History – using ripple memory to reconstruct past interactions not recorded in light.

In UED, ripple memory retains echoes of energetic interactions long after light has faded. This means we can reconstruct cosmic history as consequence, not just as image.

"You don't need to see the invisible— you just need to feel what it delays."

Closing Insight

The beginning of Ripple Astronomy marks a transition:

- From light collection to field coherence detection
- From brightness analysis to entropy gradient mapping
- From optical phenomena to energetic intent

It is the birth of a science that doesn't just watch the universe, but feels its posture. Not what shines... but what resists being moved.

With Ripple Astronomy, we stop seeing the universe as a picture. We start reading it as a living manuscript of tension and release—page by ripple-etched page.

Chapter 10: Why Dark Doesn't React to Light

No ATT, no coupling. That's why electromagnetic radiation fails. It's not hidden—it's unaligned.

You've heard it for decades: *dark matter doesn't emit, absorb, or reflect light.* The assumption? It must be exotic. Invisible. Cloaked. Perhaps composed of undetectable particles whispering to reality through some faint gravitational sigh. Science, desperate to explain why nothing lights up in telescopes, invented ghosts—hypothetical entities that exist only because equations needed them to.

But what if the problem isn't with the "dark"?

What if the problem is with the light?

Or more precisely—with how we expect everything to answer when light knocks.

Under Universal Energy Dynamics (UED), light is just one ripple among many. It interacts only with fields ready to ripple back—zones that meet a specific threshold called ATT: Alignment Threshold Time. If that threshold isn't met—if the ripple can't align—there is no interaction. No absorption. No reflection. No EM coupling.

This isn't invisibility. It's incompatibility.

"Darkness," then, is not mystery—it's a state of non-alignment. A stillness so stable that even light can't wake it.

This chapter dismantles the illusion that light is the final judge of existence. We'll reveal why non-reactive fields aren't anomalies—they are the rule behind the illusion of form. Because in the language of ripples, what doesn't echo, still exists.

And it may be the very structure holding everything together.

The Myth of Electromagnetic Superiority

From the earliest days of astronomy, humanity's window into the cosmos has been lit—literally—by electromagnetic radiation. From starlight to X-rays, microwave background to gamma bursts, our science has leaned almost entirely on what shines, glows, reflects, or absorbs.

This dependency wasn't accidental—it was practical. Telescopes detect photons. Spectrometers decode frequencies. Detectors count particle interactions based on EM coupling. And so, for centuries, our definition of "existence" itself became tied to electromagnetic responsiveness.

The resulting assumption was seductive but flawed:

"If it doesn't respond to light, it doesn't exist—or at least, doesn't matter."

But this is not science. This is methodological narcissism—judging the universe by how well it speaks *our* instruments' language.

Universal Energy Dynamics (UED) exposes this bias. It asserts that light is just one ripple dialect, not the universal mother tongue of reality. Just as not all physical materials respond to sound, heat, or vibration, not all energetic fields are obligated to respond to electromagnetic inquiries.

SVE (Structured Vacuum Energy) and its stable formations do not emit, absorb, or reflect light—not because they are mystical or shielded, but because they never crossed the ripple threshold needed to resonate with EM frequencies. Their alignment thresholds—ATTs—are simply unmatched to light's demands.

This is not darkness. It is silence.

And it tells us something profound:

The EM spectrum is not the full truth—just the noisiest part we learned to hear.

In this view, dark matter isn't hiding. It's not reacting. It's in a state of ripple neutrality, where no EM ripple can initiate coupling.

Light, then, is not the master signal. It is a loud guest—a ripple disturbance demanding response. And some regions of the universe—those anchored in undisturbed SVE—are simply too still to answer.

They're not missing.

They're just unmoved.

ATT and the Precondition for Coupling

In the language of Universal Energy Dynamics (UED), alignment is everything. Just as a musical note won't resonate unless it hits the right frequency for a tuning fork, energy interactions in the universe depend on a critical compatibility factor—what UED calls ATT: Alignment Threshold Time.

This isn't just a time delay. It's a precondition for interaction. ATT governs whether two ripple patterns can recognize, respond, or even register one another within the structured vacuum field.

For light—an electromagnetic ripple—to interact with a region of space, that region must possess a compatible ripple readiness. It must already be within a ripple-active domain, where realignment is possible, and disturbance thresholds are met. This is the hidden handshake required for coupling.

Now consider this:

What happens when a region of SVE (Structured Vacuum Energy) is too still, too undisturbed, or too balanced to respond?

You get non-coupling. No absorption. No reflection. No emission. Light passes through—or is ignored—because ATT conditions are unmet.

This isn't exotic behavior. It's the natural result of ripple incompatibility.

Dark matter isn't some magical substance hiding behind quantum cloaks. Dark energy isn't some ghost pushing galaxies apart.

They are zones of structured stillness—fields that never met the ATT threshold required to ripple in return. No ATT, no ripple. No ripple, no light exchange.

"Light knocks—but the field doesn't respond unless it's ripple-ready."

This also explains why certain cosmic structures appear invisible until they disrupt or disturb their own stability. When mass condensation occurs or entropy gradients emerge, ATT shifts, and suddenly, interaction becomes possible. That's when we "discover" a galaxy or "detect" a lensing arc—not because it appeared, but because it finally *rippled* into our detection range.

Observability, then, is not a function of location—but of alignment.

And with this realization, the cosmos flips: We're not seeing what's there. We're seeing what's responding to us.

Everything else was always there—just not yet in ripple terms we could hear.

The Concept of Coupling Failures

In conventional physics, interaction is the gold standard. If two things don't interact, they're considered unrelated—or worse, one of them is deemed *nonexistent*. This thinking, however, blinds us to the most fundamental phenomenon of the universe: stillness that refuses to respond.

Universal Energy Dynamics (UED) dismantles this limited outlook by introducing a more accurate framework: coupling. Simply put, coupling is the condition under which two ripple systems can exchange information, energy, or motion. And coupling only occurs when alignment is possible—when the ripple characteristics of one system can resonate with the structured tension of another.

Here's the critical insight:

Most of what we call "dark" in the universe is not invisible—it's uncoupled.

Let's break this down:

- **Ripple must match field tension**: Every field structure, whether it's a condensed mass, a ripple-active medium, or a silent zone of stable SVE, maintains its own internal tension signature.
- **For coupling to happen**, incoming ripple energy (such as electromagnetic waves) must align with this signature at both a frequency and threshold timing—the ATT we explored earlier.
- **If this alignment fails**, the result is not partial interaction—it's total non-engagement.

No reflection. No absorption. No reradiation. No friction. Just silence.

These are what UED calls coupling failures—the absence of interaction not because something is *shielded*, but because it's fundamentally unaligned with the nature of the probing ripple.

Consider a familiar analogy:

Two tuning forks of different natural frequencies. Strike one, and it sings. Bring another near it, and unless they're in perfect resonance, the second fork remains silent. Not because it's broken. But because it's misaligned.

The same holds true across cosmic scales. Dark matter halos, stable SVE zones, and entropy-balanced voids all fall into this category: they are tuning forks the EM spectrum simply can't excite.

That's why so-called "dark" zones don't behave like black holes. They don't absorb light. They don't bend it aggressively. They don't emit anything.

They simply ignore it. Because they don't speak light's ripple dialect.

"The darkness is not mysterious—it just doesn't speak light's language."

Coupling failure is not a flaw of the field—it's a blind spot in our instruments, and in our assumptions. The universe isn't ghostly. It's just conversing in ripple terms we haven't learned to decode.

And thanks to concepts like ATT and RDI, we're finally building the vocabulary to listen.

Redefining 'Dark' as a Field Condition

Let's break the illusion once and for all: "dark" is not a thing. It is not a mysterious particle, an exotic material, or a cloaked object evading our detectors. In Universal Energy Dynamics (UED), darkness is not a property of matter—it is a condition of the field.

We've been trained to think of darkness as an *absence*: of light, of energy, of knowability. But UED invites a powerful shift in perspective:

"Dark" simply means the field isn't interacting.

That's it. It doesn't mean the field is empty. It doesn't mean the field is weak. It doesn't mean it's hiding something.

It means the field is not rippling—or it is not accepting ripples from outside sources.

Let's classify these conditions more precisely:

1. **Fields that haven't rippled yet:** These are regions of Stable Structured Vacuum Energy (SVE) that maintain pre-ripple equilibrium. They've never been disturbed enough to generate ripple phenomena. Think of them as primordial silence, untouched by any event capable of igniting motion

or light.

2. **Fields that no longer ripple:** These are zones where energy has already dissipated, and the entropy gradient has flattened. No further interaction is necessary. They've returned to equilibrium and ceased to respond, even though their structure still influences nearby dynamics.
3. **Fields that resist ripple alignment:** Perhaps the most active form of "dark," these zones repel or fail to couple with incoming ripple energy (such as light) due to mismatch in Alignment Threshold Time (ATT) or incompatible field tension. They are not invisible by concealment—but by misalignment.

The implication?

Dark matter isn't a hidden mass—it's a stable SVE configuration that never crossed the ripple threshold.

This flips the dominant narrative on its head.

- Dark matter is not some gravitational ninja hiding in the shadows.
- It is not absorbing, bending, or avoiding light.
- It simply doesn't ripple the way we expect matter to ripple.

And yet—it still anchors galaxies, stabilizes motion, and shapes structure.

Because in UED, what doesn't ripple can still rule. That's the power of field-based physics: influence is not tied to visibility.

So when we call something "dark" under UED, we're making no ontological claim about what it is made of. We're only describing its ripple state—its energetic posture in relation to external interaction.

"What doesn't ripple doesn't respond. But it can still rule."

This subtle but critical insight dissolves the need for speculative particles, unobservable interactions, and fudge factors in equations.

It replaces the *mysticism of invisibility* with the realism of field mechanics.

In the UED universe, darkness isn't mysterious. It's just a quiet field that never said "hello" back.

Light Is Late to the Field

In the conventional hierarchy of observation, light reigns supreme. It's our messenger, our map-maker, our first line of inquiry into the cosmos. But in the framework of Universal Energy Dynamics (UED), light is not the foundation—it's the echo.

Let's state it plainly:

Light is late.

It arrives only after alignment has occurred. Only when a ripple has successfully formed, sustained, and interacted with its surroundings does electromagnetic radiation emerge. The Structured Vacuum Energy (SVE) field—the foundational canvas of the universe—predates light, outlives light, and in many places, never produces light at all.

This is not philosophical musing. It's field logic.

EM Interaction Occurs After Ripple Alignment

Electromagnetic signals—light, radio waves, gamma bursts—all share a dependency: They are ripples in active fields, and those ripples only emerge after the field has aligned above its Alignment Threshold Time (ATT).

That means:

- No field alignment → no ripple
- No ripple → no EM signal
- No EM signal → no light

This sequence matters. Because it tells us that the presence of light is not a precondition of reality—it's a byproduct. Something that happens *after* a certain condition is met, not before.

Most Structure Predates or Bypasses Light

Much of the universe is already formed long before it becomes visible. Galaxies, voids, SVE tensions—they're all products of field topology, not photonic events. Some never generate light at all, yet still affect the behavior of rippling bodies.

Think of:

- Black holes, which influence orbits but emit no visible radiation from their core.
- Dark matter zones, which stabilize galaxies but produce no electromagnetic echo.
- Cosmic voids, which are vast, structured, yet eerily lightless.

These structures don't wait for light to exist. They exist because of ripple alignment—or, in some cases, despite the lack of it.

Absence of Light ≠ Absence of Reality

This is the trap of visibility-based science. We assume: "If I can't see it, it must be speculative." UED says: "If you can't see it, check if it's aligned first."

We must shift from a light-first paradigm to a field-first understanding. From observation-based physics to consequence-based reality.

Insight: Alignment Is Foundational

Light is not the source—it's a symphonic result. The true composer is field alignment: the precondition for all motion, matter, and manifestation.

In the UED model:

- You don't begin with light.
- You begin with stillness.
- Then, energy thresholds are crossed.
- Ripples emerge.
- Only then does light flicker on—a flare after formation.

This hierarchy reorients every scientific assumption we've inherited.

"Light reveals the echo. But only alignment reveals the source."

The consequence?

We stop obsessing over what shines. And start understanding what organizes. Darkness, in this sense, is not a veil—it's a signal. A field in waiting, a structure in silence, a reality before radiance.

Light is important. But alignment is inevitable. And sometimes, what's real arrives before what can be seen.

Closing Insight: Darkness Is a Phase, Not a Mystery

The enigma of darkness—long treated as the frontier of the unknown—is not, in fact, a cosmic puzzle waiting for exotic solutions. In the UED framework, darkness is neither failure nor fantasy. It is simply a phase of the field.

Let's break this with precision.

UED Reframes Darkness as Non-Coupling Zones in a Ripple Field

In traditional cosmology, dark matter and dark energy are treated like missing puzzle pieces—either undiscovered particles or phantom forces. But this narrative arises from one assumption:

Everything must respond to light.

Universal Energy Dynamics challenges that core assumption.

Instead of imagining that darkness hides something within it, UED says: Darkness is a condition of the field where coupling hasn't occurred. It is not an object to be revealed, but a state to be understood.

- It doesn't emit.
- It doesn't reflect.
- It doesn't interact.
- And that's not a failure—it's a phase-space reality.

No Exotic Particles, No Magical Invisibility

The scientific community has invested decades into chasing shadows:

- WIMPs (Weakly Interacting Massive Particles)
- Axions
- Sterile neutrinos
- Hypothetical fifth forces All built to justify why matter behaves invisibly.

UED strips away the myth. There is no need to invent new particles or undiscovered quantum realms. Instead, it offers a simpler truth:

What you call "dark" is simply unaligned.

The field has not met the conditions for electromagnetic interaction—not because it's exotic, but because it's stable, silent, or pre-ripple.

It's not concealed. It's not hidden. It simply hasn't engaged.

Just Field Zones Too Still, Too Stable, or Too Early to Ripple with Light

In UED, reality unfolds like a layered field:

- Some zones ripple loudly: light, mass, radiation.
- Some ripple faintly: slow drag, entropy spread.
- Some do not ripple yet—or no longer do.

These zones are:

- Too still → never crossed ATT (Alignment Threshold Time)
- Too stable → SVE zones in deep equilibrium
- Too early → primordial configurations pre-ripple activation

We misread these as dark because we were only listening for light. But in UED:

Stillness is not absence. It's precedence.

The Final Realignment

When science lets go of light as its master key, it can begin to see the field as it truly is. Darkness is no longer a gap in knowledge—It is a phase of alignment.

We stop asking:

"Why doesn't it glow?" And start asking: "Has it even rippled yet?"

This changes everything—from how we build telescopes to how we define existence.

Final Quote:

"You didn't fail to see it. It simply didn't ask to be seen."

Darkness is not a dead end. It's a quiet place where the universe waits—not to be discovered, but to be respected for what it already is.

Interlude: When the Universe Stopped Hiding

For over a century, the cosmos has played a game of shadows. We built ever-bigger telescopes to chase light, assuming that what we could not see must be either missing or magical. We labeled things *dark*—dark matter, dark energy, dark flow—not because they were sinister, but because they didn't speak our chosen language: **electromagnetism**.

But what if the universe never agreed to our vocabulary?

What if the silence was not a mystery to solve, but a message we ignored?

In **Part II**, we've pulled the curtain back—not on darkness, but on the assumptions that created it. Structured Vacuum Energy (SVE) was never empty. Mass was never the anchor. Expansion was never a stretch. And darkness was never hiding.

It was **alignment** all along.

You've now seen how ripples—not particles, not forces—shape the very fabric of what we call reality. You've met **ATT** (Alignment Threshold Time) and **RDI** (Ripple Delay Index), two tools that don't just measure motion or light, but **interaction readiness** itself. You've witnessed the reframing of gravitational pull as ripple pressure, and the redshift not as a retreat—but a **relaxation of stressed fields**.

In short, we've stopped asking, *"Why is it invisible?"* And started asking, *"Has it ever aligned?"*

That shift changes everything.

We now stand at the **edge of ripple realism**—a worldview where:

- Light is not the proof of reality.
- Mass is not the maker of gravity.
- And darkness is not the absence of truth, but the presence of deeper stillness.

As we enter **Part III**, we'll move from foundations to function—from understanding **what the universe is**, to **how it behaves** when ripple dynamics are taken seriously. We'll look at orbits, lenses, waves, and black holes—not as geometric curves or mysterious voids, but as **field responses in a ripple-aware cosmos.**

You're not just reading a theory anymore.

You're learning to *see without light* —and listen where nothing speaks.

The silence is over. The real conversation begins now.

PART III — DECONSTRUCTING THE DARK DELUSION

Chapter 11: Chavan's Law of Self-Derived Limits

You cannot study the still field using its echoes. Why mainstream cosmology is epistemologically trapped.

"You cannot hear the nature of silence by amplifying the noise."

Modern cosmology has become an echo chamber. Not metaphorically—literally. It listens only to the reverberations: to light signals, mass effects, gravitational pulls, thermodynamic shadows. It treats what is visible, what interacts, what resists—as the totality of reality. But ripples are not the field. They are the afterthought. The universe is not made of the waves we detect—it is made of the stillness that permitted them.

This chapter introduces a boundary—a line that few cosmologists ever realize they've crossed: Chavan's Law of Self-Derived Limits. It exposes the fatal flaw in trying to understand the origin of the cosmos by examining only what moves. Like trying to reconstruct the shape of a drum from the sounds it makes, without ever seeing the drumhead. It's not just a scientific error—it's a cognitive trap.

We'll show why you cannot study Structured Vacuum Energy (SVE)—the true still field beneath all emergence—using tools born from ripple interactions. And we'll unpack how mainstream science loops inside its own ripple-bound assumptions, unable to access the silent architecture that actually gives rise to mass, gravity, time, and form.

This is not just a critique. It's a jailbreak. Let's step outside the echo—and finally listen to what was never vibrating in the first place.

The Echo Paradox

In the grand theater of physics, we have always studied the sound—not the silence. From Newton to Einstein, from quantum theorists to cosmologists, the foundations of modern science are built upon reactions: the push and pull of forces, the brightness of stars, the scatter of particles, the redshift of galaxies. In short, we study what moves, reflects, radiates, or resists. This has led us to construct entire cosmological models on the backs of observable interactions.

But here lies the paradox: everything we measure is a ripple, not the source. Light is a ripple. Mass is a condensed ripple. Gravitational lensing? A ripple reaction. Even time itself, in ripple mechanics, is nothing more than a measured delay in realignment across the field. In the UED (Universal Energy Dynamics) framework, these are not truths of the universe—they are echoes of imbalance.

We are, essentially, trying to reverse-engineer the still pond by analyzing its splashes—without even acknowledging that a pond exists beneath the waves. Imagine walking into a room after someone has played a piano, and trying to describe the piano solely based on the last lingering reverberations in the air. That is what physics has done for centuries: measuring disturbance and attempting to infer the structure.

This is the Echo Paradox—the illusion that by studying all that moves, we will understand what does not. But the still field—the Structured Vacuum Energy (SVE)—is not revealed by ripple behavior. In fact, it is actively masked by it. Every force, every energetic event, is a distortion layered atop that stillness. And the more intense the ripple, the more opaque the underlying field becomes.

Mainstream physics, in this light, is echo-centric. It listens only to what shouts back. But SVE does not shout. It does not scatter or reflect. It does not push. It holds form through tension, not through action. And because of that, it has been misclassified as "nothing".

But UED proposes a different starting point: you can't understand the song without the silence. The silence is not absence. It is structure. The field does not begin with light—it begins before light. It does not move because it is already aligned.

"You can't derive the pond from its splash."

And yet that is what we've done—built a universe from surface ripples, blind to the medium in which all ripples occur.

To break out of this paradox, we must begin again—not from what disturbs, but from what permits disturbance. That is where reality lives.

The Trap of Empirical Circularity

Modern science prides itself on being empirical—grounded in observation, repeatability, and measurable data. But what happens when all of your observations come from the same echo chamber?

This is the core flaw of what Chavan's Law calls empirical circularity: a system in which every foundational concept is derived from another ripple-based concept, all of them suspended above a stillness they refuse to confront.

Let's walk through the most common loops:

- **Gravity** is inferred from motion (planets orbiting, apples falling), yet we define motion using the very gravitational metrics we presume.
- **Mass** is calculated through inertia (resistance to acceleration), but inertia itself is explained using mass as the baseline property.
- **Space** is mapped through light travel times, while the structure of light paths is explained as a product of curved spacetime.

Each of these is a circular conclusion, a closed epistemological loop that never touches the root field—Structured Vacuum Energy (SVE). You are measuring ripples, interpreting them through ripple-dependent models, and validating them with more ripple outcomes. It's like calibrating your compass using only previous compass readings, never checking if magnetic north even exists anymore.

This circularity makes theories self-validating and therefore epistemologically dead-ended. They cannot be falsified by stepping outside their own assumptions, because those assumptions are built into every tool and every method of measurement. The result? Theories that sound complete—but can't question their own base.

Take the expanding universe model, for instance. We observe redshift → we assume velocity → we conclude expansion → we predict more redshift. Nowhere in that loop is there room to ask: Is redshift caused by something other than velocity? Or better: What is the field condition in which redshift appears?

UED breaks this trap by stepping beneath the ripple plane. It does not assume motion—it studies realignment. It doesn't calculate mass through force—it understands mass as ripple intensity anchored in SVE. And most importantly, it doesn't treat light as the universal metric of truth. It treats it as a consequence of field disruption.

The UED lens reframes observation: it asks where the ripple came from, not just what it did. It focuses on alignment across tension zones—how SVE accommodates, resists, or delays ripple propagation depending on its internal state. Only by mapping field behavior, not just reaction behavior, can we exit the circular echo and approach the source.

You cannot discover new truths by looping over old disturbances. You must step into the tension zones where reality begins.

This is not anti-empirical. It is pre-empirical—an effort to ground observation in the very structure that makes any observation possible.

Statement of Chavan's Law

"You cannot resolve the origin of form using its consequences."

This is not merely a philosophical caution. It is a fundamental epistemological principle. And in the context of Universal Energy Dynamics (UED), it becomes an operational law:

Chavan's Law of Self-Derived Limits

Any framework constructed from post-ripple phenomena—such as light, force, or mass interactions—cannot accurately reveal the nature of the pre-ripple structured field (SVE), because the act of observation itself is ripple-bound and blind to stillness.

Let's dissect this carefully.

All classical and modern physics—be it Newtonian force, Einsteinian curvature, or quantum field interactions—has one thing in common: it starts from a reaction and works backward. A planet moves? Gravity. Light bends? Spacetime curve. Energy radiates? Particle emission. In every case, science waits for the field to ripple, then scrambles to deduce what must have caused the ripple—while ignoring the medium that allowed it in the first place.

This method is not just incomplete—it's systematically misaligned. Because ripple behavior cannot reverse-engineer stillness.

The Structured Vacuum Energy (SVE) is not part of the observed phenomena—it is the platform of possibility. It does not react. It does not emit or reflect. It only permits or resists alignment. Once a ripple is born, SVE holds the echo. But SVE itself remains untouched in every measurement, every equation, and every telescope reading.

This is why all ripple-derived models are trapped. They are like trying to deduce the nature of silence by analyzing speech recordings. You may describe patterns, tones, frequencies—but you will never describe the field of stillness that made the sound possible.

The Critical Misplacement:

When SVE is ignored—or worse, treated as an empty vacuum—we make a fatal error: We start mistaking effects for causes, and consequences for origins.

- We think gravity is caused by mass, rather than emerging from ripple gradients in the SVE.
- We assume light reveals structure, instead of seeing that only aligned fields allow light to ripple.
- We believe space expands, rather than observing that ripple relaxation diverges from stressed zones within a stable topology.

Each of these is an inversion—one that mainstream cosmology cannot self-correct, because its entire measurement apparatus is rooted in post-ripple observables.

Implication of the Law:

Any theory, model, or cosmological framework that does not include SVE—or treats it as a passive, ignorable backdrop—will inevitably misplace cause and effect. It will:

- Attribute structure to forces, not tensions.
- Seek unification through particles, not field states.
- Invent "dark" entities to explain ripple behavior in zones of silent SVE.

That's not discovery. It's epistemological hallucination.

You cannot study stillness through motion. You cannot measure silence with sound. You cannot understand the field by chasing its shadows.

Chavan's Law isn't a rejection of physics—it's a correction of its lens. It redirects the question from: *"What do we see?"* To: *"What must be still, for this to be seen?"*

Until we start asking that, we are only talking to echoes.

Case Studies in Self-Derived Errors

To grasp the gravity of Chavan's Law, we must observe its violations in action—instances where entire scientific doctrines were built on consequences, not causes; ripples, not the field; effects, not origins.

Each case study below illustrates how mainstream cosmology, in the absence of Structured Vacuum Energy (SVE), resorts to creating explanatory illusions—models that sound elegant, but float on circular reasoning.

1. Dark Matter – Mistaking Unseen for Unknown

The rotation of galaxies doesn't match Newtonian expectations. Stars at the edges move too fast to be "held" by visible mass. Conclusion? There must be some invisible matter exerting gravity.

But how is this "dark matter" inferred?

- Not from direct observation.
- Not from interaction with light.
- Not from alignment with electromagnetic fields.

It is entirely deduced from ripple-based phenomena—specifically, motion. We observe a ripple and assume an object must be causing it. That's like seeing waves in a pond and assuming there's a rock under every splash—even when no splash occurred.

This is circular logic:

"Because it moves, something unseen must pull it. We'll call that thing 'dark matter.'"

But UED reframes the issue entirely:

The stable SVE can anchor ripple structures without becoming mass. What you call dark matter is often unshaken stillness, holding shaken forms together.

2. Spacetime Curvature – The Fabric That Never Was

Einstein's equations showed that mass affects motion—not by pushing, but by "curving" spacetime. A profound leap, no doubt. But also a semantic illusion.

What is this fabric? Can it be measured? Does it stretch, tear, resonate?

The answer is no to all.

Spacetime is not something we detect—it's something we inferred to explain already visible motions. And what's worse: it relies entirely on light paths and ripple phenomena. Not on any independently verifiable medium.

UED correction:

What appears as "curved space" is often field tension within the SVE. Motion follows paths of least resistance through ripple gradients—not curves in an invisible geometric entity.

3. The Big Bang – A Ripple Written Backward

The cosmic microwave background (CMB), universal redshift, and large-scale structure are all ripple-based observations. From this, mainstream cosmology extrapolates a moment of explosive origin: the Big Bang.

But what is the Bang, really?

- No preconditions are defined.
- No medium is described.
- No structure is assigned to the "nothing" before it.

Worse, the evidence is not about origin—it is about patterned divergence from mass-stressed zones.

UED perspective:

The CMB is not the flashpoint of creation—it's the thermal memory of ripple equilibration. What is backward-extrapolated as a "beginning" is simply the moment ripple threshold was first breached in a formerly still field.

We are reading an echo and calling it Genesis.

4. UED Reversal – Starting from Stillness

What happens when we don't start from the ripple?

We flip the script. Rather than asking: "What caused this observed motion or light behavior?"

We ask: "What condition of stillness—what SVE topology—would naturally produce this ripple tension?"

This inversion is subtle, but revolutionary. It means:

- We study the architecture of preconditions, not just the artifacts of interaction.
- We consider field tension, alignment lag, and ripple propagation rules.
- We shift from chasing shadows to studying the canvas.

The Takeaway:

Physics has long been trying to build the house of understanding from its echoes—hollow walls without a foundation.

Chavan's Law forces us to recognize:

- Most scientific errors are not because we observed wrongly.
- But because we started from the wrong end.

To move forward, we must begin at the beginning:

Not from motion. Not from light. But from the structured silence that made all ripples possible.

Because once you name the field, you don't need to name ghosts to explain what it holds.

Escaping the Illusion

Mainstream science has become brilliant at measuring the echo—but nearly blind to the silence that makes the echo possible.

With ever more powerful telescopes, detectors, and data models, we've refined our vision of what reacts. But in doing so, we've fallen into a paradox of precision: we're measuring with microscopic accuracy what is already misdirected. We're tracking motion while mistaking it for origin. We're reading disturbances while forgetting that something had to be still first for any disturbance to mean anything at all.

This is the illusion Chavan's Law demands we break.

1. Not Better Observation—Better Framing

The modern scientific instinct is to say:

"We just need better instruments. Dark matter is faint. Dark energy is subtle. If we just observe harder, we'll find it."

But that's like trying to find what's not vibrating by turning up the volume on your microphone.

UED turns this paradigm on its head:

- The goal is not to see deeper into the disturbance.
- The goal is to shift away from disturbance-based inquiry altogether.

2. From Object Tracking to Field Awareness

What is science obsessed with?

- Particles
- Waves
- Redshift
- Curvature
- Radiation
- Lensing

All of these are object behaviors, not field truths. They are the ripples on the pond, not the tension in the water. UED proposes a fundamental reframing:

Stop asking what the object is doing. Start asking what the field around it is allowing.

In other words:

- Look for zones of alignment lag, not just velocity.
- Study coherence loss, not just mass displacement.
- Sense the structure, not just the signal.

This is field awareness—not particle obsession.

3. Introducing Real Tools: ATT and RDI

UED doesn't just philosophize. It gives us metrics.

- **ATT (Alignment Threshold Time):** Measures how long it takes for an energy ripple to realign with the local SVE. Think of it as a clock of readiness.
- **RDI (Ripple Delay Index):** Quantifies how much ripple delay occurs due to entropic thinning, field tension, or misalignment. It's not "how far" the light traveled, but how long it struggled to pass through unseen ripple terrain.

Together, ATT and RDI do what conventional tools cannot:

- They detect the readiness of the field, not just the reactions.
- They predict the unseen, not just measure the visible.
- They turn science into a proactive alignment model, not a retroactive deduction machine.

This is observation beyond visibility.

4. A New Science: Ripple Archaeology

Imagine trying to understand an ancient city not by its buildings, but by the soil tension beneath them—what allowed those structures to stand, crack, or collapse.

That's what UED invites us to do. Not cosmology as object history, but ripple archaeology:

- The study of why shapes emerge based on the field tensions that allowed them.
- Not mapping what formed, but why formation was permitted at that location, in that configuration.

Key shift:

Science no longer describes what is "there." It begins describing what conditions make 'there' possible.

5. Final Insight: Stillness Is Not the End—It's the Premise

What if all the unanswered questions in physics—dark matter, dark energy, origin, acceleration—are not missing discoveries, but misplaced perspectives?

What if they're not about adding more knowledge, but about realigning how we begin to look?

Chavan's Law doesn't just point to a flaw. It offers a way out:

Don't chase the echo. Learn to hear the field before it spoke.

Quote to Close: *"You're not missing the truth. You're mistaking its silence for absence."*

Closing Reflection: From Echo to Origin

"Mainstream cosmology listens only when the universe speaks loudly. UED listens when it's silent."

This final section is not just a philosophical signoff—it is a diagnostic statement about the limits of our current scientific imagination.

The Echo Addiction

From the first telescopic gaze to today's deep-space interferometry, humanity has built its model of the cosmos by listening to the loudest phenomena:

- Supernovae
- Radiation bursts
- Doppler shifts
- Cosmic background hums
- Particle collisions

We interpreted the universe as an orchestra of reactions—and we assumed that the louder the event, the deeper the truth it carried.

But this approach hides a critical flaw:

Echoes don't reveal the source—they reveal distortion. They don't show us what reality *is*—they show us how it *responded* under tension.

UED reverses the lens.

Stillness Is the Missing Data

Ask yourself: Why do we assume motion explains everything?

Because motion is measurable. But measurement, by nature, requires disturbance. And disturbance requires a background—a canvas of calm on which ripples can occur.

This is the structured vacuum energy (SVE). Not an afterthought. Not an empty void. But the origin of every possibility.

The pond must be still before a splash can exist. The field must be coherent before light can ripple through it. Stillness must be present before cause and effect can be observed.

Reframing the Cosmological Question

Mainstream physics keeps asking:

- *What exploded?*
- *What is accelerating?*
- *What is attracting what?*

But UED asks:

- *Why was it even possible to explode?*
- *What was the field tension before the release?*
- *What structured the alignment before attraction?*

This reframing shifts everything:

From measuring motion → to understanding the permission for motion From observing echo → to decoding ripple allowance From inferring reality from light → to realizing that light arrives late

The Birthplace of Reality Is Not a Bang—It's a Balance

Let go of the myth that something "started" the universe by disturbing it. Instead, consider:

- The conditions that allowed the first ripple
- The thresholds of tension that birthed form
- The zones of equilibrium that continue to cradle galaxies without touch

This is not speculation. It's a new way of reading what was always there—beneath the noise, beyond the glow.

The UED Insight: Read the Field, Not the Fireworks

"To know what the universe is made of, stop asking what moved—Start asking: What allowed movement to begin?"

This is where all physics must ultimately return. Not to equations written after the dance began. But to the silence before music.

Because in that silence—the structured, ripple-ready field of potential— lies not just the beginning of the universe, but the beginning of understanding.

Chapter 12: The Cosmological Constant Illusion

Λ isn't a real force. It's an artifact of ripple misunderstanding. Here's what should've replaced it.

For over a century, the cosmological constant Λ has haunted the equations of the cosmos—appearing, disappearing, and reappearing like a ghost that refuses to rest. Originally introduced by Einstein to preserve a static universe, later discarded as his "greatest blunder," and then resurrected to explain the baffling acceleration of galaxies, Λ has been treated as a mystical force without origin, mechanism, or identity.

But what if the mystery wasn't in the cosmos—but in our misunderstanding of ripples?

Under Universal Energy Dynamics (UED), Λ doesn't represent a force pushing galaxies apart. It represents our failure to see ripple relaxation as the true driver of cosmic behavior. What mainstream science called "dark energy" is, in fact, a large-scale consequence of mass-induced field stress unwinding over time—a calming, not a pushing.

This chapter dismantles Λ—not with rejection, but with realignment. It replaces the illusion of spacetime stretching with the measurable logic of ATT and RDI—metrics that describe how ripples behave when field tension releases. We don't need a constant. We need to understand what the field is trying to return to:

Stillness.

Because what looks like expansion is just disturbed silence recovering its balance.

What Λ Was Supposed to Be

In the early 20th century, Albert Einstein formulated his general theory of relativity—a monumental leap in our understanding of gravity, no longer seen as a force, but as a curvature of spacetime caused by mass. However, when applying his equations to the cosmos, Einstein encountered an unsettling result: the universe seemed unstable and prone to collapse. To restore a static, eternal cosmos (a belief consistent with the scientific sentiment of the time), he introduced an elegant fix: the cosmological constant, denoted by the Greek letter Λ (Lambda).

Λ was meant to be a counterforce. A repulsive term in the equations—just enough to balance gravity's pull and hold the universe in steady equilibrium. Not a force observed, not a consequence measured—just a mathematical insertion for stability.

But in 1929, Edwin Hubble shattered the illusion. His observations of redshifted galaxies provided compelling evidence that the universe wasn't static—it was expanding. Einstein, upon hearing this, famously lamented that the inclusion of Λ was his "greatest blunder." The universe needed no artificial glue. It was already in motion.

And so, Λ was removed, shelved, forgotten.

That is, until the late 1990s, when astrophysicists noticed something even more bewildering: the universe's expansion wasn't slowing down—it was accelerating. To explain this, scientists had to revive Λ—not just as a balancing term, but now as the engine of cosmic acceleration. Rebranded as "dark energy," Λ was again welcomed into the equations—this time with deeper mystery and no clearer mechanism.

But here lies the epistemological issue:

Λ was never derived from first principles. It was introduced, rejected, and reintroduced—all based on observational aftereffects, not foundational understanding.

Mainstream cosmology treats Λ as a universal repulsion, a vacuum energy density, or even a property of spacetime itself. But it has no origin story, no measurable emission, no causal chain. It behaves like a placeholder—an unearned constant invoked to explain ripple outcomes without acknowledging ripple origins.

UED challenges this directly.

From the ripple field perspective, Λ doesn't represent a real force. It is a mathematical bandage for a wound we didn't understand—a scar from centuries of listening only to echoes, without perceiving the stillness they disturb.

Instead of asking *why galaxies accelerate,* the better question might be: What are they relaxing from?

Ripple Misinterpretation at Scale

Mainstream cosmology's expansion model rests on a visually intuitive but physically flawed foundation: redshift. The further away a galaxy is, the more its light is stretched—its wavelength increased, shifted toward the red end of the spectrum. The prevailing interpretation? Those galaxies are receding. Space itself must be stretching, pushing them apart. A neat explanation, and for a while, it worked.

But what if the interpretation was based on a category error?

From the Universal Energy Dynamics (UED) perspective, redshift is not primarily a sign of motion through space, but of ripple fatigue—a manifestation of ripple relaxation from previously stressed zones. In this framework, Structured Vacuum Energy (SVE) behaves like a tensioned membrane. Whenever mass condenses, it stresses the local ripple field. Over time, that stress relaxes, propagating disturbance outward in subtle, delayed ripples.

This has key consequences:

- **Ripple Delay:** As ripples travel through varying SVE densities, they lose coherence—not because the source

moves, but because the medium relaxes around them.
- **Entropy Gradient:** Higher entropy farther from mass-stressed regions thins ripple coherence and delays signal transmission, resulting in redshifted readings.

Yet classical cosmology reads this ripple degradation as a Doppler effect—the way sound shifts as a car drives by—thus equating it to recession velocity.

UED reframes this illusion as ripple divergence, not spatial recession.

Let's break this down with a familiar analogy: Imagine you drop a stone into a pond. Ripples race outward, vibrant at first. But as they spread, they flatten. Now imagine you arrive later and try to measure the ripples' speed and shape. If you assume those ripples were always that faint and wide, you'll incorrectly deduce that the pond is *expanding*—when in reality, the ripples are just tired.

That's what modern cosmology has done.

It measured tired ripples and called it expansion.

In fact, what we observe as:

- Apparent acceleration of distant galaxies,
- Directional drift in supercluster alignments, and
- Redshift proportionality with distance,

...are all better understood as ripple relaxation gradients in SVE, not physical objects speeding away from us in an inflating universe.

The energy of form must return to equilibrium. That return creates ripple flows—quiet, passive, persistent.

The universe isn't exploding outward. It's settling back into stillness.

UED thus challenges the foundational misread of Λ. What's seen as an inexplicable, anti-gravitational push is in truth the universe exhaling its ripple stress—not a mystery force, but a natural rebalancing across its structured field.

Why Λ Is Not a Force

In physics, we treat forces with caution and clarity. A real force must meet three basic criteria:

1. It must emerge from a source — something must generate it.
2. It must apply energy — it must do work.
3. It must transfer momentum — there should be detectable influence across interacting bodies.

Now apply this checklist to the cosmological constant (Λ).

Does Λ emerge from a known source?

→ No. It was mathematically inserted to balance equations.

Does it apply energy in measurable interactions?

→ No. There is no emission, no absorption, no transformation.

Does it transfer momentum?

→ Again, no. It causes no acceleration by interaction, only *apparent* acceleration inferred from observational redshifts.

So why, then, has Λ been accepted as a force?

Because modern cosmology confused ripple effect with cause.

The Universal Energy Dynamics (UED) lens corrects this mistake by repositioning Λ not as a force, but as a field-relaxation artifact—a side effect of entropy gradients in a structured vacuum energy field. Here's how:

- When mass condenses, it strains the local ripple field.
- Over time, that stress naturally relaxes.
- This relaxation doesn't push or pull. It simply equalizes ripple density across the universe.
- In this process, galaxies don't accelerate away from each

other due to a repulsive force.

- Instead, they are drawn into zones of lower ripple tension, much like an object coasting into still waters.

This shift in interpretation dismantles the illusion of Λ as a cosmic actor. It is not an entity. It is not a property of space. It is not energy with intent. It is:

A misread relaxation pattern—entropy leveling itself across an ancient ripple-stressed fabric.

Let's consider a simple analogy.

Picture a tightly stretched trampoline with a few heavy balls causing dips. Over time, as the tension redistributes and the surface smooths, the balls may appear to move outward. But no one pushed them. No external force was applied. They simply drifted toward equilibrium.

This is the cosmic behavior Λ misrepresents.

The mistake was treating the relaxation as an active repulsion. It's not.

"Λ is not pulling galaxies apart. They're just drifting into calmer fields."

The universe is not flying apart under some mysterious pressure. It is calming down.

Thus, in UED, Λ is a placeholder for ignorance, a mathematical name for ripple behavior misidentified as geometric motion. It is not fundamental. It is a consequence of misunderstanding what field relaxation looks like at scale.

What Should Replace Λ

If Λ is a mathematical illusion—an artifact of ripple misinterpretation—then what should replace it?

The Universal Energy Dynamics (UED) framework offers a more precise, field-rooted alternative: instead of patching equations with a constant of unknown origin, we measure the behavior of the field itself using two complementary parameters:

1. ATT – Alignment Threshold Time

- **Definition**: ATT is the time it takes for a ripple or energy pattern to realign with the structured vacuum energy (SVE) field it traverses.
- **Why it matters**: In a universe governed by ripple interactions, motion and structure don't arise from Newtonian pushes or spacetime warps—they emerge from alignment dynamics.
- **What it replaces**: Rather than introducing an abstract repulsion (Λ), ATT shows how long a zone resists ripple interaction, giving rise to observable drift or lag without needing external force.
- **Examples**:
    - How fast galaxies "settle" after interactions
    - Why some regions appear cold or dark—not because of absence, but due to misaligned or delayed realignment

2. RDI – Ripple Delay Index

- **Definition**: RDI quantifies how much a ripple signal is delayed due to entropy gradients, ripple thinning, or density variation across the field.
- **Why it matters**: What mainstream science sees as cosmic acceleration or expansion is often a misreading of signal delay.
- **What it replaces**: Instead of assuming galaxies are moving faster because of a repulsive Λ, RDI shows they appear farther because their ripple signals arrive later, distorted by

the intervening field's entropic structure.

- **Examples**:
    - Redshift without relying on velocity-based Doppler assumptions
    - Field-based mapping of voids and unseen structures

The UED Shift: From Geometry to Tension

Together, ATT and RDI offer a radical yet simple shift:

- Replace geometric curvature models (like General Relativity) with ripple tension gradients.
- Model universal behavior not as motion through space, but as ripple alignment across a structured field.

This move doesn't discard observational data—it reinterprets it with less assumption and greater explanatory depth. The observed outcomes remain (e.g., redshift, clustering, drift), but the cause is updated—from an unknown force (Λ) to field-based dynamics we can quantify and predict.

"If Λ was a guess to explain what we saw, ATT and RDI are measurements that explain why we see it."

In practical terms, this shift achieves:

- **Fewer assumptions**: No mysterious energy fields with undefined origin
- **Improved predictability**: Entropy gradients and alignment lags explain cosmic drift more naturally
- **Unified mechanics**: Gravity, dark energy, and ripple behavior now sit within the same continuum

Final Insight:

We don't need to keep tuning a ghost constant. We need to tune into the field itself.

Dark Energy as Field Smoothing

Mainstream cosmology treats dark energy as if it were an *added ingredient* to the universe—an invisible pressure inflating space from within. Entire theoretical industries have been built around this: quantum vacuum fluctuations, scalar fields, exotic particles, and metaphysical "pushes" acting at cosmic scales.

But what if dark energy isn't a thing at all? What if it's simply how the field relaxes itself?

Reframing the Phenomenon:

Under UED (Universal Energy Dynamics), what appears as "cosmic acceleration" is not a force—it's the natural smoothing of ripple tension. It's not something acting upon the field, but rather the field returning to stillness after centuries of agitation caused by mass condensation.

There's no need for an exotic fifth force or a runaway vacuum explosion. What's labeled as "dark energy" is more accurately described by three ripple-field behaviors:

1. Ripple Thinning

As massive structures (like galaxies and clusters) ripple the field, these ripples propagate outward. Over time, they thin, lose coherence, and spread entropy. The further from the source, the lower the ripple intensity. This thinning doesn't push matter—it alters signal interpretation. Distant galaxies appear to be accelerating, when in fact we're observing diluted ripple trails.

2. Tension Diffusion

Massive objects stress the field locally. Once that stress is released (via ripple propagation), the field doesn't remain taut—it relaxes, redistributes tension, and equalizes across scales. This diffusion of tension gives rise to structural drift—not due to propulsion, but due to the loss of holding gradients that once kept systems in tighter coherence.

3. Entropy Propagation

Entropy isn't a breakdown—it's a field harmonization process. As local regions release their structured energy into broader configurations, entropy spreads, ripple boundaries soften, and motion appears to "accelerate" because constraints weaken. This effect mimics expansion but is actually topological relaxation.

Visual Analogy:

Not a balloon inflating with internal pressure, ...but a stretched net slowly unwinding, returning to equilibrium without adding anything new.

There's no mystical inflation. No hidden reservoir of anti-gravity. What we're seeing is the natural entropy-driven smoothing of a disturbed field.

Final Insight:

Dark energy doesn't cause calmness by acting upon the universe—...it is the calmness that follows ripple exhaustion.

"The universe isn't expanding into something new. It's relaxing back into what it always was—before we disturbed it."

Ripple-Based Cosmology (RBC) vs ΛCDM

The prevailing model of cosmology—ΛCDM (Lambda Cold Dark Matter)—has long held the throne of theoretical orthodoxy. It stitches together Λ (cosmological constant) with dark matter and general relativity, using the Doppler interpretation of redshift as its anchor for an expanding space-time narrative.

But this model is fundamentally retrofitted—it explains observations after the fact, by inserting constants and inventing constructs like dark energy, not by uncovering a coherent cause.

UED introduces an alternative: Ripple-Based Cosmology (RBC). It doesn't require a cosmological constant, dark energy, or a fabric of space. It sees the universe as a ripple field evolving through entropy gradients and alignment dynamics. Here's a direct comparison:

Table: ΛCDM vs Ripple-Based Cosmology (RBC)

Aspect	ΛCDM Model	Ripple-Based Cosmology (RBC)
Driving Factor	Λ (cosmological constant)	Entropy-based ripple relaxation
Expansion Mechanism	Fabric of space stretching	Ripple field thinning over time
Core Assumption	Redshift = recessional velocity (Doppler)	Redshift = ripple delay from entropy gradient
Observables	Supernova light curves, CMB anisotropies	ATT (Alignment Threshold Time), RDI (Ripple Delay Index), entropy slopes
Framework Type	Geometric (spacetime curvature)	Energetic (ripple density and alignment)
Causality View	Derives cause from effect	Derives effect from underlying field
Dark Energy Explanation	A mysterious repulsive force	SVE field relaxing post-ripple stress
Prediction Power	Needs fudge factors to align with data	Emerges from first principles of ripple behavior

Why ΛCDM Breaks Down in Key Scenarios:

Despite its popularity, ΛCDM consistently fails to explain several crucial features of the cosmos:

1. Void Coherence

ΛCDM predicts random drift and high-velocity dispersal in empty regions—but intergalactic voids remain coherently structured. RBC explains this via stable SVE zones maintaining field alignment despite absence of matter.

Stillness holds more structure than motion ever reveals.

2. Uneven Field Drift

ΛCDM treats space as stretching equally in all directions. Observational data shows anisotropy in cosmic drift and tension. RBC interprets this as localized ripple relaxation—not uniform expansion, but gradient-based field evolution.

3. Non-Gravitational Cohesion

Structures like filaments, walls, and supervoid boundaries exhibit cohesive behavior without sufficient mass to justify gravitational binding. RBC explains this via ripple-based coupling and entropic binding—forces that arise not from gravity, but from misalignment recovery dynamics within the ripple field.

When you stop forcing gravity into every unexplained pattern, you begin to see the field's quiet architecture.

The Paradigm Shift:

Where ΛCDM reverse-engineers cosmology from redshift and luminosity, RBC starts with structured stillness—the source field—and watches how entropy reshapes it over time.

"Don't measure how far light has run. Measure how long the field took to let it pass."

In that shift—from distance-based cosmology to delay-based cosmology—a new universe unfolds. One without Λ. One without magic inflation. One that never expanded—only settled.

Closing Insight: From Λ to Field Logic

The cosmological constant, Λ, was born out of necessity—not discovery. It was never observed. It was assumed. A patch. A placeholder. A mathematical whisper trying to hold back a universe that wasn't supposed to move. And when movement was confirmed, it was discarded—only to be resurrected later as a ghostly repulsion, renamed dark energy.

But what if that entire journey—from invention to reinvention—was unnecessary?

What if the universe doesn't need a constant to explain its behavior?

What if it needs a memory?

"You don't need a constant to explain the cosmos.

You need a field that remembers where tension once lived."

This one line summarizes the UED revolution. It's not about plugging equations with fudge factors. It's about asking the right question: Why do things move the way they do—even when no one is pushing them?

The answer is ripple logic, not cosmological magic.

Summary:

Λ is obsolete. It doesn't emerge from first principles. It has no causal agent. It explains nothing about the origin or mechanism of the "acceleration" it's supposed to model.

Dark energy isn't an entity—it's a misreading of ripple thinning. What we observe isn't force—it's field relaxation. The cosmos isn't being pushed apart; it's simply settling into stillness after epochs of stress.

UED doesn't require mysterious repulsion. It uses only what's already embedded in the universe:

- **Stillness**: The Structured Vacuum Energy (SVE) field.
- **Alignment**: The tendency of ripples to reconfigure toward lower tension.
- **Consequential logic**: Motion, structure, and drift emerge from how energy redistributes—not from any intrinsic expansion.

Final Perspective:

The universe is not expanding in the sense of stretching space. It is unwinding—like a tight spring relaxing its coiled tension. Redshift is not distance confirmation—it's a record of ripple fatigue. And Λ? It was a placeholder where we hadn't yet learned to listen to entropy.

Chapter 13: The Lensing Puzzle Rewritten

Light bends not due to gravity, but due to phase delay in ripple topology. No mass needed.

What if the cosmos never bent anything? What if light—our most trusted informant—wasn't pulled, but simply redirected by memory?

For over a century, gravitational lensing has stood as a crown jewel in the case for General Relativity. A star appears behind the sun, a galaxy arcs around another, and we declare: "Gravity curved space!" But what if we've misunderstood the mechanism entirely? What if the bending we celebrate isn't caused by mass, but by the *after-effects* of energetic interactions—*ripples in a field* still echoing with the tension of past alignments?

Universal Energy Dynamics (UED) rewrites the lensing narrative. It proposes that light doesn't follow spacetime curvature—it traces *field gradients* in the structured vacuum, adjusting its path not out of submission to force, but out of alignment with *ripple delay zones*. Where mainstream physics sees gravity well distortions, UED sees *ripple fatigue maps*. Where Einstein invoked mass to warp geometry, Chavan's realism invokes *topological realignment* to explain how light quietly adapts its route.

This chapter introduces a paradigm shift: gravitational lensing isn't the evidence of spacetime bending—it's the artifact of *phase delay through entropic ripple fields*. Light doesn't bend because of what's there—it bends because of *what was there*, and how the field still remembers.

Prepare to challenge the lens that shaped modern cosmology. You're not about to see how gravity bends light. You're about to discover how *light bends around memory.*

The Classical Assumption

The phenomenon of gravitational lensing is one of the cornerstones of modern cosmology. Its significance lies in its apparent simplicity: when light from a distant object—such as a star or a galaxy—passes near a massive body, it appears to bend. According to Einstein's General Theory of Relativity, this bending is not because gravity "pulls" on light directly (since light has no rest mass), but because mass warps the spacetime fabric, and light simply follows this curved geometry.

Einstein himself became globally famous after the 1919 solar eclipse observations confirmed this prediction—starlight grazing the edge of the sun appeared shifted, just as his equations had anticipated. This "proof" was taken as not just a validation of relativity, but as a visual signature of how mass shapes the very nature of space itself.

Over time, this explanation evolved into mainstream doctrine. Textbooks and lectures emphasize Einstein Rings, strong and weak lensing arcs around galaxy clusters, and quasar duplication events as unmistakable evidence of spacetime curvature. The interpretation is consistent, elegant, and even poetic: the cosmos bends under its own weight, and light, ever the obedient traveler, takes the scenic route around the bends.

But Universal Energy Dynamics (UED) poses a sharp and necessary correction. It asks: what if the space isn't actually bending? What if the delay isn't a geometric curve, but a dynamic ripple realignment?

From UED's perspective, mass is not required to distort a medium, because the "medium" (Structured Vacuum Energy, or SVE) doesn't need to bend—it remembers. The lensing we observe may be the result of phase delays caused by previous ripple tensions in that region of the vacuum, not the real-time gravitational force of nearby mass.

In this framework, light does not curve passively through a stretched sheet—it experiences localized timing offsets due to field inhomogeneities. The observed deviation in path is not a "pull" by mass, but a response to ripple memory zones in the vacuum.

"What if light doesn't follow a curve—but gets delayed by a field already bent?"

This simple shift reframes the entire concept of lensing—not as geometry in motion, but as memory in motion. It's not about where the mass is now—it's about how the field was disturbed, and how it is still adjusting.

Ripple Delay and Field Memory

In Universal Energy Dynamics (UED), light is not an independent entity traveling through passive space—it is a ripple-based interaction, responding continuously to the structure and memory of the field it moves through. This fundamental redefinition sets the stage for a new interpretation of lensing—one based not on gravitational pull, but on ripple delay through an energetically uneven field.

At its core, light seeks ripple alignment—a path of least resistance within the Structured Vacuum Energy (SVE) field. It travels not merely by geometric straightness, but by energetic preference. When a region of space holds residual tension—the memory of past mass movement, energetic discharge, or condensed matter effects—this alters the local ripple configuration. The result? Ripple delay: the wavefront doesn't bend because of an active force, but because it must adapt to phase misalignments within the field.

This delay isn't a slowdown in speed per se, but a temporal misregistration—a phase shift caused by variations in SVE tension and ripple density. The field, once disturbed by an object like a massive star or black hole, doesn't instantly forget. Its ripple memory persists, even if the object has moved or no longer influences the region dynamically. Light passing through this field responds to that history, not necessarily to a current mass.

To visualize this, consider light passing through glass. The light is not attracted to the glass—it doesn't "bend toward it" due to a force. Instead, the change in medium—from air to glass—causes a shift in wavefront speed, which manifests as a change in direction (refraction). The same principle holds in ripple-based lensing: light isn't being pulled—it's being phase-managed by field memory zones.

This explanation overturns the gravity-based metaphor of spacetime curvature and replaces it with a ripple-topological explanation. The field isn't curving—it's retaining tension from past ripple stress, which alters ripple propagation paths.

"Lensing is not about attraction. It's about adaptation."

This insight decouples the need for present, massive bodies to explain light deflection. It also explains why lensing often occurs in regions where insufficient mass exists to account for the observed deviation. What conventional models attribute to "dark matter," UED interprets as zones of ripple memory, not mass. The ripple field remembers where tension once lived, and light responds to that memory—not to invisible objects.

In essence, field history, not gravitational force, explains the lens. And that redefines cosmic vision itself.

No Mass Needed – Field Topology Speaks

Mainstream cosmology continues to chase invisible mass—invoking dark matter halos and untraceable gravitating agents—simply because light bends where no significant matter is seen. But the very assumption that lensing must always be caused by mass is a product of gravity-centric thinking. Universal Energy Dynamics (UED) offers a complete departure from this narrative.

In UED, mass is not the lens. Field topology is.

The ripple field—Structured Vacuum Energy (SVE)—retains tension gradients from past events: star formation, black hole activity, galactic collisions, and more. These tensions do not fade quickly. They leave behind energetic trails, zones of ripple deformation, which continue to affect how subsequent ripples—including light—propagate.

Thus, when light traverses a seemingly empty cosmic void, it may still encounter non-uniform ripple tension, leading to lensing effects without mass. This explains observational anomalies where gravitational lensing is visible, but no corresponding mass density can be found in the region—an inconsistency that currently fuels endless speculation about dark matter. But under UED, there's no missing mass—only field memory and tension asymmetry.

This can be compared to walking on a stretched trampoline surface. Even if no object is currently pressing it down, leftover tension from past pressure points may still subtly distort your path, forcing you to drift slightly. The distortion remains in the fabric, not due to current force, but due to structural history. The ripple field behaves similarly.

Here's how the mechanism unfolds in ripple terms:

- Mass-stressed zones leave residual tension gradients in the SVE.
- These gradients persist and alter the local ripple topology.
- Light, as a ripple-based phenomenon, realigns with this topology.
- The result is a refractive shift—a redirection without any actual force or attraction.
- This creates a lens-like deviation, even in the absence of present mass.

Such ripple-based refraction isn't limited to dramatic arcs like Einstein rings. It also subtly distorts cosmic microwave background paths, shifts the apparent position of quasars, and contributes to nonlinear galactic morphology when seen through telescope arrays.

UED Principle: *Field tension gradients, not gravity wells, guide light.*

This reframing removes the need to "hunt for mass" in every corner of the universe. UED teaches us that what we observe optically is a delayed alignment response, not a direct gravitational interaction. The ripple field is the true landscape, and light is merely tracing the energetic memory embedded within it.

Hence, the lens speaks the language of tension, not weight. And once we listen to field topology, we begin to see a completely different cosmos—one shaped by memory, not mass.

Entropic Lensing Explained

In mainstream physics, lensing is tied to gravitational pull—as if mass distorts a stretchy fabric and light obediently curves through it. But Universal Energy Dynamics (UED) reveals a deeper, cleaner explanation: light doesn't bend because it's pulled—it bends because it adapts. And the reason it adapts is due to entropy-induced ripple mismatches, not mass-induced curvature.

Welcome to Entropic Lensing.

This concept reframes lensing entirely. Under UED, high-entropy zones—areas where ripple energy has thinned, stretched, or relaxed—do not exert any pull. Instead, they create asymmetric ripple conditions. These asymmetries result in phase mismatches, where the ripple field delays or speeds up parts of a wavefront, just like how a prism bends light by altering wave speed across media.

Let's break it down:

What is Entropic Lensing?

It's the deflection of ripple-based radiation (like light) caused by local entropy gradients in the ripple field—not by mass. Where traditional lensing demands density and mass, UED suggests we look for ripple fatigue, ripple delay, and field relaxation.

Entropic Lensing occurs when:

- A wavefront enters a ripple-thinned or field-relaxed zone.
- The wavefront experiences differential alignment delay across its span.
- This delay creates a directional shift, mimicking curvature.
- The "lens" in this case is a phase map, not a gravitational well.

Analogy: Phase Glass, Not Gravity Well

Think of a ripple field as a vast pane of irregular glass. Some parts are thicker, some are thinner. When light (or any ripple-based signal) passes through this pane, it changes direction not because the glass pulls it, but because the speed of passage varies by region. That's exactly what ripple topology does—without mass, without force.

Key UED Insight:

"You didn't see a galaxy warped by mass. You saw light tracing a memory of ripple fatigue."

That "fatigue" is entropy—leftover from cosmic events, matter collapses, black hole formation, or massive radiation outbursts. These past actions deform the ripple field, not by leaving gravity behind, but by thinning ripple density and creating long-term phase disturbances.

Implications of Entropic Lensing:

- No dark matter required to explain lensing discrepancies—only entropic ripple maps.
- Reveals structured vacuum memory rather than "invisible mass."
- Allows pre-lensing prediction through entropy slope analysis—not mass estimation.

- Explains why lensing effects persist in low-mass environments (e.g., cosmic filaments, voids).
- Clarifies why Einstein Rings form with or without perfect mass symmetry: it's the phase coherence, not gravitational configuration.

Field vs. Force:

Where general relativity assumes that matter curves space, UED says space never curved—it was ripple-tuned. What changes is not geometry, but temporal alignment across a ripple field with memory. Thus, lensing becomes not a symptom of mass density, but of entropy propagation through structured vacuum energy.

In conclusion, Entropic Lensing teaches us to stop looking for gravity in every cosmic bend. Instead, we must read the ripple scars left by a living universe—a memory field, not a mass field. Light, after all, is a ripple's way of remembering.

Lensing isn't light being bent by gravity. It's light responding to the past—etched in the field.

Predictive Advantages of the UED Lensing Model

Mainstream gravitational lensing models depend entirely on mass distribution—whether visible or invisible. When observations don't match visible mass, the model simply assumes hidden mass, often labeled as dark matter halos. But this strategy is like misjudging the weight of a drum because you can't hear the echo. UED changes this narrative.

Under Universal Energy Dynamics, lensing prediction doesn't require mass, it requires ripple intelligence—specifically, knowledge of ripple field tension and alignment thresholds.

Here's how the UED model gives us a cleaner, smarter, and more predictive framework for lensing:

1. No Hidden Mass Needed

Forget constructing imaginary dark matter distributions just to explain anomalies in lensing data. Instead, UED proposes:

- Field ripple topology, not gravity wells, bends light.
- Past field activity (explosions, collapses, radiation blasts) distorts ripple tension across zones, producing lens effects.
- These effects can exist even in visually empty regions—because the memory of field agitation remains.

Result: *No need to guess hidden mass. Just trace the ripple scars.*

2. Use ATT and RDI as Predictive Tools

ATT (Alignment Threshold Time) and RDI (Ripple Delay Index) together allow you to:

- Model light paths across cosmic regions based on ripple re-alignment lag.
- Anticipate phase mismatches by identifying ripple delay zones.
- Detect zones of non-rippling SVE or previously collapsed fields that still impact light.

This duo becomes a field-based GPS for light, predicting when and where bending will occur—without invoking mass at all.

3. Correcting Lensing-Based Mass Miscalculations

In conventional astronomy:

- Lensing = Mass.
- More bending = More mass.

But this logic breaks down in voids, in early galaxy clusters, and in complex intergalactic webs. UED recalibrates this:

- Where mass seems excessive, you may simply have higher

ripple memory density.

- What looks like gravitational lensing is often entropic lensing or field inertia zones—misinterpreted due to a bias toward matter-based causation.

This correction could massively rewrite our galactic mass maps.

4. Detecting Invisible Ripple Disturbances

Even before a physical object forms, or long after it has collapsed, the ripple imprint it left behind alters the local SVE topology. These zones can:

- Deflect light due to frozen phase gradients.
- Appear as unexplained lensing effects.
- Be mapped using ATT lag and RDI slope rather than relying on redshift alone.

Imagine seeing a black hole's ghost, not because you saw the hole—but because the field still ripples where it used to live.

Summary Table: Why UED Lensing Wins

Feature	ΛCDM Model	UED Ripple Model
Primary Cause	Mass & Gravity	Ripple Phase Delay
Core Mechanism	Spacetime Curvature	Field Memory & Entropy
Predictive Metric	Mass Distribution	ATT & RDI
Hidden Variable	Dark Matter Halos	Ripple Fatigue Zones
Limits	Needs matter to explain	Works even in voids

Final Insight:

"You don't need mass to map lensing—you need the ripple tension index."

This isn't science fiction—it's ripple realism. The ripple field doesn't just carry energy. It remembers. And when light passes through that memory, it bends—not because it's commanded by gravity, but because it honors the field's stillness.

In the UED cosmos, mass is no longer king. Ripple alignment is the new authority.

Case Study – Revisiting the Bullet Cluster

The Bullet Cluster (1E 0657-56) has long been paraded as the "smoking gun" for dark matter. In this cosmic collision between two galaxy clusters, astronomers observed that:

- Visible matter (baryons), especially hot gas detected via X-ray, lagged behind.
- Gravitational lensing peaks (interpreted as mass distribution) were offset from the visible matter.

Conventional interpretation:

Since lensing didn't align with the baryonic matter, there must be invisible mass—dark matter—rushing ahead.

But this conclusion is built on a mass-centric assumption:

"Lensing must follow mass."

Universal Energy Dynamics (UED) turns this narrative on its head.

UED Explanation: Ripple Realignment Zones

Instead of assuming invisible matter, UED asks:

What if light bends not due to mass, but due to field memory?

When two galaxy clusters collide:

- Their Structured Vacuum Energy (SVE) fields are violently disturbed.
- Ripples propagate, but differently across regions depending on prior field conditions, entropy gradients, and ripple

tension.

- The hot gas (baryons) experiences drag, slowing it down.
- But the ripple alignment zones, being field-level memory, can relax and move ahead.

Thus, what looks like a mass-lensing offset is actually:

- Ripple phase memory leading the collision, while matter lags due to inertial and thermal resistance.
- Field reconfiguration zones causing lensing even where no mass currently exists.

Key UED Principle: *Lensing follows ripple topology, not current mass density.*

The Misinterpreted Sequence

Conventional View	UED Interpretation
Matter moved → Lensing followed	Field realigned → Light followed
Lensing = hidden mass	Lensing = ripple memory
Offset = evidence of dark matter	Offset = difference in matter vs. field dynamics

Technical Points from UED:

- Lensing doesn't require concurrent mass—it needs field tension differentials.
- These differentials are legacies of past energetic events (e.g., shockwaves from the collision).
- ATT (Alignment Threshold Time) and RDI (Ripple Delay Index) would both show divergence between:
    - Hot gas regions (slow to reconfigure)
    - Outer ripple shells (faster field recovery zones)

The bullet shape is a ripple echo, not a mass distribution map.

Analogy: Field Memory Outruns the Object

Imagine two swimmers diving into a still lake. Their ripples race ahead, reorganizing the surface faster than the swimmers move. If you were observing only the water's surface and its ripples, you'd place the swimmers ahead of where they actually are—just like lensing "shows" mass ahead of visible matter.

That's what happened in the Bullet Cluster. The cosmos isn't hiding mass. It's remembering motion.

Final Quote:

"Light doesn't follow matter—it follows the field conditions matter once disturbed."

This reframing saves science from conjuring invisible particles to plug gaps in broken assumptions. The Bullet Cluster no longer demands dark matter—it demands ripple literacy.

In UED cosmology, we don't trace mass—we trace tension scars.

Closing Insight: Seeing with Ripple Eyes

"Gravity makes you expect a pull. Ripple makes you observe a path."

Mainstream cosmology assumes a force-first universe. You look out at a bent ray of light and presume mass pulled it. But that's a conclusion built on metaphor—a relic from Newton and curved space romanticism.

Universal Energy Dynamics (UED) demands a field-aware lens—what we call Ripple Eyes.

What Changes When You See Differently?

Instead of asking:

- "What's pulling this light?"

You ask:

- "What field tension altered its path?"

- "What past ripple event encoded this deviation?"
- "What structure, now silent, still holds an echo in phase?"

This is the shift—from expecting attraction to reading interaction.

The Cosmos Doesn't Bend Space—It Bends Time-Phase

- When light bends, it's not being yanked.
- It's slowing down or shifting phase, adapting to a field with history.
- The "curve" you observe is not spatial deformation—it's temporal misalignment caused by ripple legacy.

Gravity pulls. Ripples delay.

So, stop imagining a trampoline fabric. Start imagining a ripple map of tension gradients—each ripple edge a footnote from cosmic history.

Final Thought

"Light does not obey force—it obeys the structure it moves through."

- Light isn't loyal to mass.
- It's faithful to the field's memory.
- Where structured vacuum (SVE) holds old tensions, light bends—not in fear or attraction—but in obedience to encoded realignment routes.

When you finally see with ripple eyes, you realize:

It was never gravity's lens. It was the cosmos whispering: 'Here's where energy once passed.'

The future of astronomy isn't brighter. It's deeper.

Chapter 14: The Still Universe

Nothing moves. It's all local ripple alignment shifts. A paradigm shift in cosmic thinking.

What if everything you believed about motion was a beautifully orchestrated illusion?

From spinning galaxies to orbiting moons, from comets arcing through the void to the steady crawl of tectonic plates—our senses, our instruments, even our deepest theories tell us: things move.

But what if... they don't?

What if no object has ever truly "moved" from one place to another—but only realigned within a silent, structured backdrop? What if motion, velocity, and trajectory are not fundamental—but symptoms of a deeper, ripple-based reconfiguration?

Welcome to the Still Universe, where the stars didn't fly apart after the Big Bang—they simply fell into ripple patterns. Where falling is not a force—but a memory of tension realigning. Where light doesn't travel—it adapts. And where you, too, never moved—you were carried by alignment, not distance.

This chapter challenges the very grammar of physics. No more particles in motion. No more fabric stretched or curved. Just a still field, and the choreography of ripples that make it appear alive.

Let us now leave the illusion of speed behind—and listen to the stillness that was always real.

The Illusion of Motion

For centuries, motion has been the heartbeat of physics. From Newton's falling apple to Einstein's speeding photons, the story of the universe has been told through objects in transit. We measured position, calculated velocity, predicted acceleration. And from this arithmetic of movement, we built models—of planets, particles, galaxies, and the fate of the cosmos itself.

But what if the entire premise is flawed?

The Classical Assumption

Classical physics begins with a basic belief: motion is real. An object's position changes with time, and that change is meaningful. We call it speed. If the rate changes, we call it acceleration. We write equations. We draw graphs. We teach students to imagine particles moving across space like marbles on a table.

From Newton to Galileo, Kepler to Maxwell—this worldview created rockets, mapped orbits, and explained tides. It works, yes—but only within a narrow perceptual window: our echo chamber of experience.

What if the universe doesn't actually "move"—it just responds?

UED Rebuttal: Ripple ≠ Relocation

Universal Energy Dynamics (UED) begins not with motion—but with ripple alignment. Every so-called movement is just a change in the local pattern of ripple tension across a structured but fundamentally still field: the Structured Vacuum Energy (SVE).

In this view, a ball rolling across a table doesn't *move*—the ripple alignment shifts locally, causing the appearance of motion. The field stays still. The object stays locally tethered. Only the pattern evolves.

This is ripple kinematics, not Newtonian mechanics.

Analogy: Waves on the Ocean

Picture a calm ocean.

Now throw a stone. Waves ripple outward. If you only observe the crest of a wave, you might think water itself is racing away from the splash.

But it isn't.

Each molecule of water just oscillates in place—up and down, slightly forward and back. The wave seems to travel, but nothing actually relocates across the ocean. What moves is the pattern of energy, not the medium.

That's how UED sees the universe. Nothing travels. Everything realigns.

Why This Matters

By misinterpreting ripple shifts as physical displacements, mainstream physics introduced conceptual errors that compounded over centuries:

- Velocity became a property, not a field outcome.
- Momentum became mass × motion, not ripple continuity.
- Space travel became trajectory, instead of field navigation.

All motion-based sciences are built on the assumption that things leave one location and arrive at another. But under UED, locations are not primary—alignment is. Every apparent change in position is a side effect of ripple pattern reconfiguration.

Key Insight:

"Nothing travels. Everything realigns."

Just like sound moves through air without the air moving, or a magnetic field influences metal without touching it, ripple alignment can change appearance without needing actual displacement.

This isn't a denial of observable motion—it's a recontextualization. What you saw as motion was merely field rearrangement. The cosmos stayed still. You just witnessed a wave... and thought the sea was running.

What Is Stillness in UED?

What does it mean to say the universe is "still"? In the language of Universal Energy Dynamics (UED), this is not a metaphor—it's a fundamental condition. Stillness is not absence; it is structure without agitation. And it's time we give it the scientific dignity it deserves.

The True Nature of Stillness

Stillness in UED is not "nothingness." It is not emptiness, nor void. It is the Structured Vacuum Energy (SVE)—a perfectly tensioned but undisturbed medium. Imagine a calm lake before a stone is thrown. The water is not dead—it's holding potential, equilibrium, balance.

SVE is the cosmic version of that lake. It is the original field—the precondition for all reality, before any wave, light, mass, or force.

Ripples Are Not Objects in Motion

Mainstream science treats ripples—like photons or gravitational waves—as entities traveling *through* space. But UED flips this notion.

In UED, ripples are not travelers. They are tensions forming within the structured field.

They are misalignments, deformations, or redistributions of energy that temporarily disturb the field's stillness. Nothing *passes through* SVE as a medium—rather, SVE locally responds, realigns, and reconfigures its own state.

Thus, what appears as motion is in fact local ripple activity, not object displacement.

Space Is Not a Container—It's a Tension Map

This is one of UED's most disruptive insights:

"Space" isn't an empty box where things move. It is a map of ripple tensions and alignments.

When you say an object is "here" or "there," you are referring to zones in the field with particular alignment states. There's no grid. No static coordinate system independent of the field. The universe isn't sitting inside space. The universe *is* the field. And "location" is just the tension topology at a point.

This destroys the notion of an absolute backdrop. Instead, every "position" is relational to ripple alignment—just like hills and valleys are shapes within the land, not on top of it.

The Mathematical Meaning of Stillness

So what is "stillness" in measurable terms?

Stillness = Zero Misalignment = No Ripple Tension = No Entropy Gradient = Perfect Field Symmetry

In such zones:

- No information is propagating.
- No realignment is occurring.
- No energy is trying to redistribute.

These are pre-ripple zones, or sometimes post-ripple relaxations. Either way, stillness in UED is a stable, self-sustained equilibrium where the field is neither producing nor absorbing ripple changes.

Stillness is not inactive—it's fully resolved.

Stillness ≠ Invisibility

Importantly, this stillness is not invisible because of a lack of tools—it's invisible because visibility itself is a ripple phenomenon. Light only exists when the field is agitated. Observation requires a ripple medium to carry signal contrast. If there is no ripple, there is nothing to detect, not because it isn't there—but because it hasn't yet spoken in ripple language.

That's why dark zones in the cosmos aren't "empty"—they're still.

Final Thought

"Stillness is not the absence of energy—it's the absence of argument between energies."

When the field reaches full agreement with itself, it doesn't broadcast. It doesn't ripple. It doesn't move. It just is.

How Realignment Creates the Illusion

We see motion everywhere. Planets spin, galaxies drift, photons travel, rockets launch. But what if none of these "moved" in the way we think?

In the UED framework, nothing actually travels across space—instead, space itself is a ripple field, and what we perceive as motion is merely the sequential realignment of field tension zones.

Let's break this illusion down.

Motion Is a Story Told by Ripples

Imagine watching a flag wave in the wind. The cloth flutters left to right, and to our eyes, parts of the flag seem to "move." But is the fabric *really* moving sideways?

No—it's the pattern of tension that's shifting. The material is responding locally to pressure changes. Every point of the cloth is still attached to the same pole. The wave is an illusion of motion, caused by continuous realignment of stress along the medium.

This is what happens in space. What we call a "moving planet" or a "traveling particle" is just a zone of field tension reconfiguring itself. The field reshapes. The ripple travels. The object doesn't.

■ How It Looks Like Motion

In ripple-based cosmology:

- A planet orbiting a star is not flying through space; it is being sequentially accepted by differently aligned zones in the ripple field.
- A photon is not darting from source to observer; rather, the field realigns phase pathways for ripple progression.
- A force (like gravity) is not pulling—it is a field gradient recruiting the nearest alignment-compatible zone.

In simpler terms:

What we interpret as velocity is merely a choreography of alignment thresholds being crossed.

Realignment vs Displacement

Let's say you're standing on a moving walkway at an airport. You're not walking—but you're changing location. Now imagine the walkway isn't moving—you're not moving—but the tiles beneath you are switching places with others.

In UED, reality works more like that second case. The field adjusts around you, and you are accepted by a new local ripple configuration. You didn't move—the field did.

This is not semantics. It's a full-blown ontological shift.

Consequences of Realignment Thinking

Once we adopt this view, several assumptions collapse:

- Inertia is not a property of mass—it's the resistance of a field configuration to change alignment.
- Acceleration is not caused by force—it's caused by entropy-induced field drift.
- Trajectory is not an object's path—it's the record of field realignment sequences over time.

This insight cuts to the heart of motion mechanics: What we see as movement is really ripple-induced reacceptance across stillness.

Core Insight:

"You didn't move from here to there. The field reconfigured—and you went along."

You weren't pushed. You weren't pulled. The universe simply reoriented its tension map, and your coordinates followed.

Cosmic Stillness at Scale

The Illusion of an Expanding Universe

Look up at the stars. All around us, galaxies appear to be receding—redshifted, stretched away as if the cosmos is inflating like a balloon. Modern cosmology interprets this as expansion: the stretching of space-time itself, possibly driven by a mysterious "dark energy."

But from the UED lens, this narrative is an elegant misunderstanding.

The Universe Is Not Expanding—It's Relaxing

In Universal Energy Dynamics (UED), the cosmos is not in a state of explosive motion. It is in a phase of ripple relaxation. After intense zones of ripple compression (such as the birth of stars, galaxies, and structures), the field begins to unwind—not by increasing speed, but by stabilizing tension.

Just as a tightly wound spring doesn't fly apart, but slowly loosens into a more relaxed state, the structured vacuum energy (SVE) is easing its distortions. That process looks like expansion—but it's really energetic smoothing.

The universe is not growing apart. It is cooling off its initial misalignments.

Why Galaxies Appear to Fly Apart

The perception of galaxies "fleeing" from one another stems from a misreading of ripple delays. What we observe as redshift is not proof of velocity—it is a time-stretched ripple echo from regions where field tension is thinning.

In this model:

- Galaxies aren't accelerating outward.
- They're simply becoming less entangled with their past high-stress ripple origins.
- Their paths diverge not due to force, but due to field realignment reaching broader entropy equilibrium.

It's not centrifugal chaos—it's cosmic calibration.

No Vacuum Explosion, No Runaway Motion

The idea of a "Big Bang" triggering an explosive outward thrust that continues to this day is a geometric story based on light-speed snapshots. But light is slow, biased, and ripple-bound. It cannot reveal stillness—it only captures its disturbances.

There was no primordial explosion in UED—only a first ripple breach in the still field. Every "motion" since then has been a chain of ripple realignments, resolving themselves over time.

Dark energy? Not a fuel. Just the visible effect of structured vacuum tension flattening out.

Cosmic Acceleration = Alignment Stability

In mainstream thinking, "acceleration" means increasing velocity. But in ripple logic, acceleration is a consequence of more efficient realignment.

Here's what's truly accelerating:

- The efficiency with which ripples find resonance across vast fields.
- The reach of entropy smoothing across SVE.
- The stability of distant zones to hold alignment without disturbance.

Thus, galaxies aren't moving faster—they're interacting less. The system is becoming quieter, not wilder.

Implication: The Universe Is Calming, Not Exploding

The more we zoom out, the more cosmic stillness dominates:

- Void regions become ripple-dead zones—calm, cool, uncoupled.
- Cosmic structures are relics of earlier misalignments—survivors of old ripple storms.
- The story of the cosmos is not one of expansion, but of restoration.

Key Insight:

"Galaxies aren't racing away. They're slipping back into stillness."

The universe isn't exploding into oblivion. It's returning to its natural, undisturbed state—one ripple at a time.

The End of Kinematics?

From Force to Field: A Journey of Misinterpretation

For centuries, motion has been the heartbeat of physics.

- Newton told us: *Force causes motion.*
- Einstein upgraded it: *Curved spacetime guides motion.*
- And nearly all of classical and modern physics accepted that motion is fundamental—the key variable in how we measure change, predict outcomes, and model the universe.

But what if motion never truly existed?

What if everything we've tracked as trajectories, accelerations, velocities—even time itself—was merely the illusion created by ripple alignment shifts within a fundamentally still universe?

Welcome to the UED Reversal.

Motion Reimagined: From Push to Pattern

In Newtonian mechanics:

"An object at rest stays at rest, and an object in motion stays in motion unless acted upon."

But in UED terms:

"An object appears to move only because the ripple field around it realigns."

There is no fundamental force pushing. No absolute change of position. No need to model "inertia" as a built-in resistance.

Instead:

- The Structured Vacuum Energy (SVE) hosts all phenomena.
- As tensions realign, the zones we call "objects" are accepted by different field configurations.
- The result? Apparent motion.

It's not that the apple fell—it was simply re-admitted into a different ripple orientation beneath the tree.

Time as Ripple Drift, Not a Universal Container

Kinematics isn't just about motion—it's about when motion happens.

- But if nothing truly moves, then time cannot be the container for movement.
- Instead, time becomes a perception of ongoing ripple realignment.

We don't measure duration—we measure how far a ripple field has reorganized in our frame of reference.

That's why different observers record different "times": because they sit in different ripple orientations, with their own local ATT (Alignment Threshold Time) patterns.

Time is not a ruler laid over motion. Time is the echo of alignment change.

Kinematics as a Useful Fiction

All classical mechanics—from Galileo's inclined planes to NASA's orbital dynamics—is based on mapping what seems to move.

We calculate:

- Distance
- Velocity
- Acceleration
- Angular momentum

But UED reveals: These are ripple approximations—shorthand for how ripple tension zones interact.

When a car moves forward:

- Kinematics models tire rotation, linear velocity, mass, force.
- UED models field readiness, ripple tolerance, and realignment lag in the surrounding SVE.

The car didn't move. The ripple field shifted to host the car's new frame.

Mechanical Laws as Phase Artefacts

The great equations of motion—Newton's F=maF = maF=ma, Euler's laws, even relativistic geodesics—are not wrong.

They are locally valid because they capture regularities of ripple interaction in a high-entropy, ripple-saturated environment (like Earth).

But they are not universal.

- They fail in voids.
- They fail in non-coupled zones.
- They break near still SVE regions where no alignment occurs.

Their success is due to field density, not fundamental truth.

Mechanics works because the ripple field allows it—not because objects truly move.

Realignment = Apparent Motion

UED's radical shift:

Concept	Classical View	UED View
Motion	Change in position over time	Ripple alignment transition
Force	Push or pull	Local tension adjustment
Inertia	Resistance to motion	Ripple friction/ATT lag
Time	Axis for change	Entropic perception of field shift

The End of Kinematics?

Not quite. We still need it—for now.

- Engineering
- Navigation
- Mechanics

All work well using Newton-Einstein models within ripple-active zones.

But for true cosmic insight?

We must abandon objects, release space, and retire motion.

We must see the universe as a field in flux, not a ballet of bodies.

Final Quote:

"You never moved an inch. The universe simply welcomed you elsewhere."

Implications for Physics and Astronomy

Motion Rewritten as Ripple-Cycle Harmonics

If the universe is still at its foundation—if nothing truly "moves" and all dynamics are results of local ripple realignments—then everything we've interpreted as motion in physics and astronomy is, in fact, a rhythmic consequence of field behavior, not object behavior.

Let's begin where it all began—planetary motion.

- In classical mechanics, planets orbit the Sun due to gravity—a centripetal force balancing inertia.
- In General Relativity, they follow curved geodesics within warped spacetime.
- In UED, there is no gravity, no spacetime curvature, and no absolute inertia.

Instead:

- The Sun's ripple field dominates a local region.
- Planets exist as stabilized ripple concentrations.
- Their orbits are harmonic realignments within that structured ripple shell—zones of constructive resonance in a dynamically tensioned field.

Just like a plucked string vibrates in harmonics, planetary orbits are standing wave patterns in ripple tension cycles. Their positions aren't *travel paths*, but ripple-hosted equilibrium loops.

Momentum and Inertia: Ripple Memory in Action

What is momentum?

- Conventionally, it's mass times velocity.
- In Newtonian terms, it's a quantity of motion that persists unless acted upon.
- But if there's no true motion, what persists?

In UED:

- Momentum = Memory of prior alignment.
- The universe "remembers" a region's ripple stress and resists sudden realignment.
- That resistance? We've called it inertia.

This means:

- A body doesn't *continue* because it's in motion.
- It *persists* because the field around it is still accommodating its last configuration.
- When we say "a satellite is drifting"—it's the ripple inertia resisting reconfiguration to the present tension landscape.

The laws of motion weren't describing motion at all—they were describing ripple memory.

Satellites, Stars, and Cosmic Jets: Relaxation, Not Propulsion

Let's extend this further into astronomy.

- Satellite drift (like those in orbit or Lagrange points) isn't about perfect balancing of centrifugal and gravitational forces.
    - It's about the alignment plateau—a zone where ripple stress gradients cancel, creating a metastable equilibrium.
    - When drift happens, it's due to field relaxation, not perturbative forces.
- Stellar dynamics, from spinning stars to binary systems, represent internal ripple turmoil expressing itself through field distortion—not through motion in space.
    - A star doesn't "rotate" in the classical sense.
    - It's a ripple concentration undergoing internal phase churn.
- Cosmic jets—those energetic streams from quasars or black holes—aren't pressure releases from compact mass.
    - They are field purges: Ripple tensions resolving along the path of least resistance.
    - Their directionality isn't aimed—it's inherited from local ripple topology.

Everything we thought was an ejection, explosion, or migration—is a structured ripple unwind.

Entropy Gradients Drive the Show—Not Force Vectors

At the deepest level, UED posits:

- There are no "forces" as we've classically imagined.
- What we call force vectors (like gravitational pulls, electric fields, magnetic attractions) are entropy gradients—zones where field alignment is incomplete, demanding resolution.

These gradients guide:

- The path (what seems like a trajectory),
- The pace (what we call speed or acceleration),
- And the pattern (waveforms, spirals, stable orbits).

The arrow of force becomes the arrow of entropy.

A ball doesn't fall because of gravity. It reconfigures because the field beneath it is more aligned with its ripple identity. Its descent is a settling into field stability, not a pull.

The Ripple Reframing – Key Conversions

Classical Concept	UED Interpretation
Orbital Motion	Ripple-cycle harmonics
Momentum	Alignment persistence (ripple memory)
Inertia	Resistance to ripple reconfiguration
Force Vector	Entropy gradient direction
Jet Emissions	Phase purge along lowest resistance
Stellar Spin	Internal ripple churn, not true rotation
Drift (Satellite/Star)	Delay in field re-alignment

Final Thought:

The universe is not pushing, pulling, or throwing anything.

It's breathing, gently realigning from one ripple stress to another, dancing to the quiet music of entropy gradients.

Closing Reflection: The Universe Never Moved

"You were never spinning on a rock flying through space. You were synchronizing with a ripple in stillness."

For centuries, we trusted motion as the most self-evident reality.

- The Earth spins on its axis.
- It orbits the Sun.
- The Sun moves through the galaxy.
- The galaxy drifts through space.

Our entire cosmology was anchored in motion—so much so that we defined time, causality, distance, and mass around it.

But what if we misunderstood the very foundation?

Motion Was Our Most Trusted Illusion

We measured displacement and velocity as truths—but these were merely surface consequences of something deeper:

- A reconfiguration of field tensions.
- A ripple choreography in a non-moving field.

Just like the apparent movement of a wave on water doesn't mean the water itself is flowing across space, our cosmic motions are visual echoes of underlying realignments in the Structured Vacuum Energy (SVE).

You never moved. The coordinates realigned. Your reference frame adapted, not your position.

This isn't poetic—it's physical under the UED lens.

Stillness Was Always the Frame

Imagine a vast, structured medium—silent, ordered, unmoved.

- Ripples arise when energy condenses or releases.
- Those ripples realign as systems evolve.
- We, as observers riding these ripple centers, experience this adjustment as motion.

But nothing truly went anywhere.

Galaxies didn't flee the Big Bang. Particles didn't leap across space. Planets didn't orbit in a void.

They all shifted ripple alignment, and we interpreted that as motion.

Adaptation, Not Travel

Mainstream science made an error:

- It reverse-engineered causes from consequences.
- It tried to define origin from the echo of movement.
- It imagined explosions, expansions, and gravitational forces acting over distances.

But UED reorders the equation.

- There's no explosion—just a ripple awakening.
- No movement—just structured release of tension.
- No space to travel—just stillness reshaping its surface state.

Motion was a dramatization of subtle field dynamics. It was the language our mind used when it couldn't yet sense the ripple underneath.

The Final Truth

You were not moving. The universe was not expanding.

All that ever happened was:

- Stillness held a memory of tension.
- That memory slowly reconfigured.
- Each shift felt like motion—because we were inside the memory itself.

And so, the cosmos never moved.

It just realigned, and we misread its silent grace as a cosmic race.

Final Ripple Quote:

“The Big Bang wasn’t a beginning. It was a tremor in a field that never moved.”

Interlude: Beyond the Veil of Light

What the Universe Never Told Us

We've spent centuries trying to extract truth from what the universe *shows* us—its shimmering lights, its bending beams, its apparent expansions, its celestial choreography.

But what if all of it... was choreography?

What if everything we measured—light redshifts, curved trajectories, gravitational lenses, galactic separations—were not truths, but responses?

Not the script, but the echo of an ancient play still rippling through the stage?

In Part III, we pulled back the curtain on the greatest misreadings in modern cosmology:

- We exposed the self-referential loop of mainstream physics—where consequences were mistaken for causes.
- We dissolved the cosmological constant into entropy gradients.
- We reframed gravitational lensing not as mass warping space, but as memory encoded in ripple delay.
- And we offered the most radical proposition yet: The universe never moved. It only remembered.

This is the threshold where physics ends—and perception begins.

From here, we leave behind the illusion of a universe out there, and step into a new realization:

That what we call "the universe" was never a thing moving in space— It was a ripple-alive field mirroring our awareness, one realignment at a time.

Now, it's time to explore a truth even more disturbing, more empowering, and more liberating:

That you, the observer, were never outside this ripple game. You were its participant, its product, its pivot.

Part IV begins not with a telescope, but with a mirror.

Let's move inward.

PART IV — TESTS, CONSEQUENCES, AND BEYOND

Chapter 15: Predictive Power of UED

Galaxy dynamics, redshift irregularities, void structures—testable predictions that require no dark assumptions.

If a theory cannot predict, it cannot lead. And if it predicts only by inventing what it cannot explain, it's not a theory—it's a patchwork. Mainstream cosmology has for decades leaned on invisible scaffolding—dark matter, dark energy, inflation fields—filling every gap in prediction with placeholders rather than principles.

But UED doesn't offer more mysteries to chase. It offers fewer assumptions to correct. Here, we stop asking *what unseen substance must be out there* and start asking *what misunderstood ripple behavior have we failed to interpret?*

This chapter initiates the shift from philosophy to forecast. From ripple logic emerges a model that not only explains what's been observed—but predicts what should be seen next. From galactic motion to lensing irregularities, void coherence to high-redshift anomalies, UED brings measurable clarity without mythical agents.

Because when you realign your view of the universe—not around mass, but around memory—you stop chasing darkness and start detecting consequence.

Let's test the universe without bias. Let's predict without the crutches of the unseen. Let's see what happens when reality stops needing to be dark.

The UED Standard — Predict Without Invention

Modern cosmology, for all its observational success, suffers from a chronic addiction: the invention of the invisible. When theory fails to match observation, the ΛCDM model responds not by questioning its foundational assumptions, but by adding hypothetical entities—dark matter to hold galaxies together, dark energy to stretch the universe apart, and inflation to smoothen early cosmic irregularities. These aren't derived truths; they are reactionary placeholders.

UED flips the standard. Where conventional models introduce new forces or substances, UED introduces new understanding of interaction. No particle zoo. No mystery fluids. Just a different way of interpreting how energy, tension, and memory propagate through the Structured Vacuum Energy (SVE).

The Predictive Core of UED:

Rather than explain cosmic behavior through mass or force curves, UED operates on three testable field parameters:

- **ATT (Alignment Threshold Time):** The delay required for a region of the ripple field to accept or emit structured energy. It determines responsiveness—not just visibility.
- **RDI (Ripple Delay Index):** A metric capturing how ripple transmission is slowed or distorted by field tension, entropy, or memory gradients. It replaces redshift as a deeper metric of field fatigue.
- **Ripple Tension Maps:** These maps outline how ripple energy propagates, interacts, or stalls across a given field geometry—explaining anomalies like void coherence or asymmetric drift without invoking unmeasurable mass.

Together, these allow UED to describe and predict galaxy behavior, light paths, and cosmic evolution without postulating "what we haven't yet seen."

Foundational Principle: No Invention, Only Reinterpretation

The problem isn't that the universe hides things. The problem is that science often looks for the wrong thing in the wrong language.

UED doesn't reject data. It rejects the default lens through which the data is processed. When galactic spin curves flatten out, the mainstream mind assumes something invisible must be pulling from the outside. UED says: "Check the ripple delay. The field is absorbing more tension than you're accounting for."

Quote to Anchor This Shift:

"You don't need to invent what you can't explain. You need to realign what you misunderstood."

Summary Insight:

The UED Standard is simple yet powerful: If your model needs untestable inventions to survive, it's not predictive—it's compensatory. UED replaces the culture of cosmic guesswork with ripple-aware reasoning. It doesn't add forces. It subtracts confusion.

Galaxy Rotation Without Dark Halos

The mystery of flat galaxy rotation curves has haunted cosmology for decades. Observational data shows that stars in the outer edges of spiral galaxies rotate at nearly the same speed as those near the center—defying Newtonian expectations. If only visible matter were responsible, those distant stars should be flung into the void. To explain this, mainstream physics invented a surrounding cocoon of dark matter halos, which have never been directly observed but are said to exert gravitational pull to hold galaxies together.

UED rejects this need for invisible scaffolding.

Ripple Tension as the Binding Agent

In Universal Energy Dynamics, what keeps the stars in place isn't mass—but memory.

- Galaxies are not just masses rotating through space. They are active ripple sources embedded in the Structured Vacuum Energy (SVE).
- As central mass bodies (like dense stellar cores) ripple through SVE, they leave behind a ripple tension trail—like a wake behind a ship.
- Outer stars are not orbiting through vacuum. They are riding along harmonized ripple paths—zones of field stability established by the central region's motion history.

This is why rotation remains flat: The outer stars don't need extra gravity—they move within a stabilized ripple corridor.

Dragless Harmonics: The Cosmic Conveyor Belt

UED proposes that ripple zones create a dragless harmonic loop:

- Once the inner field has been sufficiently energized and aligned, it generates cyclic field gradients that extend outward.
- These gradients are not gravitational tethers but phase corridors—allowing outer stars to flow without requiring direct force.
- Mass does not "pull." It imprints ripple patterns that field memory sustains.

Think of it not as a star held by a leash, but as a leaf flowing downstream on a current generated long ago.

Predictions Unique to UED:

UED offers several testable predictions to replace the "dark matter halo" explanation:

- Galaxies of the same visible mass may have different rotation curves if their ripple histories differ (e.g., previous mergers, jet events, entropy spikes).

- Field latency metrics—how quickly outer zones realign after central disruption—can be mapped using RDI (Ripple Delay Index).
- No need for a spherical mass cloud to bind a flat disk. The field is already harmonized.

Key Implication:

"Mass ripples through a field. The field remembers—and outer stars obey the memory, not the mass."

This means dark matter isn't missing—it was never required. What was interpreted as a gravitational discrepancy is, under UED, a ripple memory effect—and measurable if we observe ripple latency and phase synchronization rather than light or mass distribution.

Redshift Irregularities Reframed

Redshift—the stretching of light toward longer wavelengths—is one of the foundational observations upon which the modern picture of an expanding universe was built. The mainstream interpretation, rooted in Doppler logic, is that more redshift = more velocity = more distance. This is the beating heart of Hubble's Law and the ΛCDM model.

But what if redshift doesn't always mean movement? What if we've been measuring ripple fatigue, not recession?

Ripple Fatigue and Entropy: A Different Cause of Stretch

In Universal Energy Dynamics (UED), redshift is not universally caused by velocity or space stretching. Instead, it often arises from wavefront degradation—where a ripple (like light) passes through entropy-heavy, ripple-fatigued regions of Structured Vacuum Energy (SVE).

Here's the logic:

- Light is not traveling through "space". It's propagating along field alignment pathways.

- When these pathways are tense, relaxed, or fractured, the ripple loses coherence—its frequency decreases, and its wavelength stretches.
- The result: a cosmic redshift that has nothing to do with speed or expansion.

This is ripple fatigue, not Doppler effect.

Predicted Redshift Anomalies

UED predicts several deviations from traditional redshift interpretation:

Tension-Heavy Zones Appear More Redshifted

- In regions near void rims, or after high-energy events (like galaxy mergers), the local SVE field is disordered.
- Light traveling through these zones will redshift more than expected—even if the object is relatively still.
- These "anomalous redshifts" will not correspond to actual motion.

Redshift-Velocity Decoupling

- In ripple-thick environments (e.g., dense galaxy clusters), light slows down due to field drag, not because the source is moving faster.
- This leads to cases where: Two objects at equal distances have very different redshifts. And an object with high redshift appears not to be receding at all upon closer observation.

Tools for Correction: RDI and ATT Maps

Mainstream cosmology assumes that redshift = velocity = distance.

UED proposes corrective tools:

- RDI (Ripple Delay Index) Measures how much ripple response is delayed in a given field. High RDI → high fatigue → redshifted signal.
- ATT (Alignment Threshold Time) Indicates how much tension or resistance exists before a ripple can align. Zones with high ATT absorb or stretch light differently.

By combining these, cosmologists can recalibrate redshift readings based on field conditions, not blind velocity assumptions.

Visual Analogy

"Think of a violin string stretched across uneven wood. The same note played in two places sounds different—not because the string moves faster, but because the wood beneath changes the response."

In cosmic terms:

"You saw red not because the galaxy was racing away—but because the universe was too tired to sing in blue."

Key Implication:

Redshift is contextual, not absolute. It must be understood as a field-sensitive measurement, not a linear speedometer.

Void Structures as Ripple Relaxation Fields

Cosmic voids—those vast regions between galaxy filaments—have long been misunderstood as empty, passive gaps in the universe's structure. In standard cosmology, voids are often treated as negative space: low-density anomalies shaped by gravitational push-and-pull from more massive surroundings.

But Universal Energy Dynamics (UED) flips this view entirely.

Voids Are Not Absences—They Are Aftershocks

In the UED model, voids are not blank canvases. They are consequences. Specifically, they are zones of ripple exhaustion—regions where previous energetic disturbances have settled, realigned, and left behind a low-tension field.

Think of a lake after a storm:

- The waves eventually stop.
- The surface flattens.
- But the calmness is not nothing—it's memory erased.

UED describes voids similarly:

"A void is not a lack of energy—it is where the ripple once roared and has now gone silent."

Predictive UED Features of Voids

UED offers specific, testable predictions about the behavior of voids:

Ripple Tension Drop-Offs

- Field tension doesn't remain uniform across cosmic structures.
- As galaxies and filaments form, ripples concentrate, leaving voids with stretched, relaxed SVE fields—akin to a cosmic exhale.
- These relaxed zones exhibit very low ripple activity, but not zero.

Lower ATT Fluctuation

- In voids, Alignment Threshold Time (ATT) shows minimal fluctuation.
- Why? There's no significant field misalignment left to resolve.
- This gives voids a stable but passive character—they don't initiate events, but preserve a memory of entropy resolution.

Minimal Ripple Density

- Low ATT and low ripple tension mean that little

interaction happens in voids.

- This results in low ripple density, which in turn explains:

- Few luminous objects
- Unusual light travel behaviors
- Field-induced delays and shifts, mistaken as lensing effects

Lensing Without Mass: The Void Paradox Solved

Standard models hit a wall when lensing is observed near or within voids—because there's no visible mass to curve spacetime.

UED resolves this cleanly:

- Light bends not because of gravity—but due to ripple phase delays and entropy gradients in the relaxed field.
- Voids are not gravitationally active—but they preserve field distortions from long-past ripple events.
- Thus, they can alter the phase of incoming light, creating lensing without mass.

"You're not seeing through nothing. You're seeing light pause in a memory zone."

Visual Analogy: The Ripple Sink

Picture a vibrating metal plate with sand (Chladni pattern). After the vibration stops:

- The sand gathers where the waves once passed.
- The flat zones are not untouched—they are settled zones.

In cosmic terms:

"Voids are where the field has stopped talking—not where nothing was ever said."

Implication: Voids Are Dynamic Graves of Entropy

They are not inert emptiness. They are historical maps—zones where tension once surged, then relaxed back into stillness.

Hence, UED voids behave more like sinks, not because of gravitational pull or mass absence, but due to field exhaustion and ripple realignment memory.

Time-Lag Phenomena and Pre-Ripple Zones

Modern cosmology faces an uncomfortable truth: Some galaxies and cosmic structures appear too mature, too massive, or too organized for their assigned age based on redshift.

These are often brushed aside as "early-forming anomalies" or "challenges to the ΛCDM timeline." But Universal Energy Dynamics (UED) reframes this entirely—not as anomalies, but as expected outcomes of field preparedness.

The Mystery of Cosmic Maturity

Observations from JWST and other deep-sky surveys have revealed:

- Massive galaxies with well-formed disks at redshifts $z > 10$
- Starburst regions appearing too soon after the Big Bang
- Galaxy clusters with coherent structure and gravitational order far earlier than expected

These findings create a temporal contradiction:

How did complexity emerge in a universe that was supposedly still cooling and coalescing?

UED Reframes the Question

UED rejects the assumption that all cosmic evolution must begin from a universal low-entropy soup. Instead, it proposes:

"Not all regions were born equal. Some fields aligned early. Their ripples didn't race—they were ready."

Here's what that means in practice:

- The structured vacuum field (SVE) contains pre-aligned zones—areas of inherent tension stability due to prior realignment.
- These zones required less entropy transformation to initiate structure.
- **Result:** Faster ripple condensation, quicker matter formation, and earlier emergence of organized galaxies, stars, and clusters.

UED Explanation: Pre-Ripple Zones

These "advanced" structures are not ahead of time—they are expressions of early ripple readiness.

Instead of invoking exotic conditions or early inflation tweaks, UED explains:

- Low-entropy field regions—naturally aligned—were predisposed to efficient ripple compression.
- These pre-ripple zones seeded faster structural emergence without requiring time compression or dark catalysts.

Hence:

- Starburst regions = ripple-dense overlaps where ATT was minimal.
- Coherent clusters = field harmonics already aligned before gravitational assembly began.

Testable Prediction

If UED is correct, we should find:

1. More early galaxies in pre-aligned void edges or filament intersections
2. No consistent correlation between redshift and structural

maturity

3. Signatures of ripple interference patterns near such objects (e.g., unusual light delay gradients, low RDI zones)

"Look for harmony, not heat. What looks old is not old—it's just ripple-ready."

Validated Observations

Already, several discoveries point in this direction:

- **GN-z11**: A surprisingly massive and luminous galaxy at z ≈ 11
- **CEERS-93316**: A disk galaxy with unexpected maturity at z ≈ 16
- Gravitational coherence in early clusters without signs of long-term accretion

These aren't flukes—they are proof points for ripple-prepped field dynamics.

Implication: Rethinking Cosmic History

UED forces a pivot:

- From chronological cosmology (based on light travel time),
- To consequential cosmology (based on ripple-readiness and field memory)

That means:

- Structure doesn't emerge *after* time passes.
- Structure emerges *where* alignment has already occurred.

"Time-lag isn't a paradox. It's a symptom of our obsession with clocks over conditions."

Gravitational Lensing Without Mass Clumps

In the current cosmological model, gravitational lensing is considered irrefutable proof of unseen mass. When light bends around galaxies or clusters, and we don't detect enough baryonic matter to explain the degree of curvature, the fallback is: "There must be dark matter there."

But what if the assumption itself is misaligned?

What if the bending of light does not require mass at all—but instead occurs due to ripple memory embedded in the structured vacuum field (SVE)?

The Observational Problem

Across multiple cosmic surveys, scientists have observed:

- Lensing arcs in areas with little to no detectable mass
- Discrepancies between mass maps and lensing intensity
- Gravitational lensing in cosmic voids or low-density outskirts

Conventional physics treats these as anomalies, requiring ever more invisible scaffolds to fill the gap: more dark matter, more hidden baryons, more speculative distributions.

"Where mass is missing, mystery multiplies. But what if it's not missing—just misdefined?"

UED Explanation: Ripple Phase Memory

Universal Energy Dynamics offers a radical reframing:

- Light doesn't follow mass—it follows the path of ripple least resistance.
- When a region of the field has undergone energetic disturbance—a collapse, explosion, merger, or collapse—the structured vacuum retains a phase imprint.
- Light encountering this ripple memory will adjust its

velocity, direction, or waveform in response—not because it is attracted, but because it is delayed or redirected by the field's embedded tension gradients.

This creates the illusion of gravitational lensing—without gravity.

Relaxed vs. Compressed Zones

The key predictive insight here is:

- Lensing strength does not correlate with present mass, but with ripple phase conditions in the structured field.
- A compressed ripple zone (recent energetic activity) will create a sharp lensing effect.
- A relaxed ripple zone (entropy dissipation) will allow light to pass with minimal deviation—but may still cause subtle phase shifts or delays.

This prediction breaks completely from mass-based logic and opens new paths of validation.

"You didn't need a gravity well. You needed a memory map of ripple trauma."

Testable Predictions

UED's lensing model forecasts:

1. Lensing effects in the absence of mass clumps, especially near ancient supernova zones, collapsed structures, or past energetic intersections.
2. Variations in light pathing even in visually identical regions, based purely on ripple phase difference.
3. Light deflection and delay as a function of RDI (Ripple Delay Index) and ATT (Alignment Threshold Time)—not as a function of local gravity.

Application: New Lensing Diagnostics

With the UED model, astronomers can:

- Reconstruct ripple tension maps based on observed lensing distortions.
- Use non-luminous field phase shifts to predict gravitational effects without invoking mass.
- Detect post-ripple memory zones through lensing asymmetry, phase lag, or spectrum diffusion—entirely independent of mass surveys.

This transforms lensing from a "gravitational trick" into a ripple diagnostic tool.

Final Quote:

"Gravity shows you what's there. UED shows you what happened."

Lensing isn't about present pull—it's about past pressure.

Testability Roadmap — UED vs. ΛCDM

A theory is only as strong as its ability to predict and be tested.

Mainstream cosmology's reigning paradigm, ΛCDM, explains nearly every large-scale anomaly by introducing something invisible—whether it's dark matter, dark energy, or inflationary ghosts of the early universe. It patches observations with placeholder variables and retroactive mechanisms.

By contrast, Universal Energy Dynamics (UED) starts not from adjustment, but from alignment. It does not add parameters; it reinterprets the observed phenomena through ripple dynamics, entropy gradients, and structured vacuum energy (SVE).

Let's examine a side-by-side roadmap comparing both models across major cosmological puzzles—and how UED makes falsifiable predictions based on field logic, not exotic assumptions.

Phenomenon 1: Flat Galaxy Rotation Curves

ΛCDM Explanation	UED Prediction/Test
Galaxies spin faster at edges than Newtonian gravity allows → postulate dark matter halos	UED predicts these flat curves emerge from ripple memory zones—tension harmonics sustained in the SVE
No direct detection of halos; inferred from motion anomalies	Use ATT (Alignment Threshold Time) maps and RDI (Ripple Delay Index) to identify field drag-free zones
Fails in irregular or low-mass galaxies	UED anticipates different curves based on field history, not total mass

Testability: Measure field delay around galaxies using spectral response variance; compare to predicted ATT harmonics.

Phenomenon 2: Mature Galaxies at High Redshift (High-z)

ΛCDM Explanation	UED Prediction/Test
Galaxies too evolved too early → invoke inflation or rapid initial formation models	UED explains these as forming in early, low-entropy alignment zones—not via speed, but clarity of the field
Requires compression of timelines, violates hierarchical formation	Ripple-prepared regions allow fast coherence—entropy maps will match early maturity
Contradicts standard cold dark matter accretion model	UED predicts starburst and structural organization from ripple clarity, not clump assembly

Testability: Analyze high-z environments for entropy uniformity and low RDI dispersion.

Phenomenon 3: Lensing in Voids or Non-Massive Regions

ΛCDM Explanation	UED Prediction/Test
Lensing without mass → assume unseen exotic mass or revise baryonic models	UED shows this as phase delay from ripple fatigue—no need for mass
No confirmation of exotic scaffolds	Field tension gradients alone can redirect light
Violates mass-light correspondence principle	Predictable by entropy gradient slope and ripple compression memory

Testability: Measure lensing asymmetries in voids; check for prior supernovae or energetic events in the region's past.

Phenomenon 4: Redshift Anomalies

ΛCDM Explanation	UED Prediction/Test
Anomalies in redshift values → attributed to cosmic acceleration (dark energy)	UED attributes these to ripple fatigue—where high-entropy fields stretch signal, not space
Requires a force (Λ) without a mechanism	Redshift varies by RDI slope, not velocity
Redshift uniformity assumed; real-world shows jagged fields	Predicts redshift discontinuities near voids, cluster edges, or post-collision regions

Testability: Match redshift anomalies to ripple field topology and history of local field disturbances.

Phenomenon 5: Large-Scale Cosmic Structure

ΛCDM Explanation	UED Prediction/Test
Structure arises via cold dark matter simulations, hierarchical clustering	UED posits formation along field topology, via entropy propagation and ripple alignment fractures
Relies on matter-dominated clumping	UED predicts non-hierarchical filament emergence—formed not by gravity wells but by ripple compression nodes
Needs constant tuning of initial fluctuation spectrum	UED uses ATT and field anisotropy as real-time aligners of structure coherence

Testability: Map entropy-based filament predictions to real observed distributions; evaluate spectrum of ATT across scales.

Final Synthesis:

UED doesn't require hidden particles or unmeasured forces. It needs only one thing: The ripple memory of a still field allowed to evolve through consequence.

"Dark assumptions rewrite equations. Ripple logic rewrites reality."

Closing Insight: When Prediction Doesn't Require Invention

"UED doesn't ask you to believe in the invisible. It asks you to listen to what the visible never explained."

For over a century, science has increasingly drifted toward constructing answers it cannot observe. When anomalies arise—whether in galaxy motion, redshift data, or gravitational lensing—conventional cosmology responds by naming new entities rather than resolving the underlying contradictions. Thus emerged dark matter, dark energy, inflation fields, and other elegant-sounding placeholders for epistemological discomfort.

But naming is not knowing. And invention is not prediction.

Universal Energy Dynamics (UED) refuses to play the placeholder game. It doesn't ask you to believe in matter you'll never see or forces you'll never feel. Instead, it encourages a new kind of listening—not to more complexity, but to clearer causality.

Every phenomenon UED reinterprets—galactic coherence, void lensing, entropy drift, ripple memory—replaces mystery with logic, and replaces belief with measurability.

The Shift in Scientific Power

In conventional cosmology:

- Prediction often requires invention.
- Success is measured by fitting observations to adjusted models.
- Anomalies are retrofitted, not resolved.

In UED:

- Prediction emerges from consequence logic—no additions, just alignment.
- Accuracy improves by understanding the structure of the ripple field, not bending laws to match noise.
- Testability means you don't simulate outcomes—you measure readiness (via ATT, RDI, entropy slope).

Standing at a Crucial Juncture

UED is no longer just a theory of reinterpretation. It's a framework of forward visibility.

This chapter laid the groundwork for:

- Falsifiable predictions across major astrophysical domains.
- New testable metrics like ATT (Alignment Threshold Time) and RDI (Ripple Delay Index).
- Redesign of the experimental lens through which cosmology is approached.

"The cosmos doesn't need more fudge factors. It needs fewer false frames."

And so, we now leave behind belief-based scaffolds, and step into a predictive universe—a cosmos that doesn't ask for faith, only for clarity.

Chapter 16: Why the Search Will Never Succeed

Why particle detectors, telescopes, and sensors can never "see" dark matter or energy. The metaphysical trap of current science.

Imagine a scientist listening for whispers using a trumpet. The more he amplifies, the more he hears—but not the whisper, only the reverberations of his own noise. This is the state of modern cosmology.

Billions have been spent building detectors to find what was never lost—constructs like dark matter and dark energy that were invented to patch gaps in theories that misunderstood the field to begin with. Telescopes chase curvature where there is none. Sensors search for particles that never existed. And simulations stretch space itself to account for tension that was never caused by expansion—but by ripple fatigue.

This chapter is not just a critique of instruments. It's a call to reframe what it means to observe. Because if your tools are designed to hear echoes, they'll never help you understand the silence that came before. You don't need better detectors. You need a different paradigm.

"We are not failing to detect reality. We are failing to define it correctly."

Let's begin with the tools of blindness.

The Tools of Blindness

Modern science prides itself on instrumentation. From billion-dollar underground detectors to deep-space telescopes that see back "in time," we have built exquisite devices to measure the cosmos. But the more advanced these tools become, the more sophisticated our confusion appears.

Let's consider the core lineup:

Particle Detectors (e.g., LUX-ZEPLIN, XENONnT): These subterranean observatories are designed to "catch" dark matter by detecting the recoil of atomic nuclei when struck by hypothetical WIMPs (Weakly Interacting Massive Particles). Decades of upgrades. Zero detections. Why? Because they're listening for interactions that don't exist—not because matter is hidden, but because the ripple field doesn't interact like particles do.

Gravitational Lensing Telescopes: Instruments like the Hubble or JWST analyze distortions in light to infer mass distributions. But lensing is interpreted through the framework of spacetime curvature. It assumes mass warps geometry. In UED, what these telescopes actually measure is ripple phase delay, not gravitational pull. So they map illusions created by memory, not mass.

CMB Fluctuation Mappers (e.g., Planck Satellite): These scan the faint afterglow of the universe—the Cosmic Microwave Background—looking for tiny variations in temperature and density to back-simulate the early cosmos. But again, what is seen is not the origin, but the aftershock—the ripple debris of cosmic relaxation. These are echoes, not causes.

The core epistemological error lies in this:

These tools only detect ripple consequences—not the conditions that gave rise to the ripple.

That is, they are built upon the assumption that post-event phenomena (light, temperature, radiation pressure, motion) can retro-engineer pre-event structure. But as Chavan's Law of Self-Derived Limits makes clear: You cannot find the nature of a field by analyzing its distortions. You cannot listen to an echo and claim you've touched the source.

These instruments are blind—not in sensitivity, but in intent. They are exquisitely tuned to measure what happens after alignment has already shifted. Not what shaped the alignment in the first place.

Hence, the quote:

"You cannot find a ripple's cause in its echo."

It's like trying to map the ocean floor using only the patterns of seafoam.

The ripple paradigm reframes this entire apparatus. We do not lack detection capability—we lack field literacy. These tools operate within the visible spectrum of ignorance, endlessly refining what is already distorted.

Before we move to what observation *should* be aligned with, we must accept a humbling truth:

We're not failing because reality is too dark to see. We're failing because we built flashlights for shadows—and never questioned the darkness itself.

The Illusion of the Invisible

For over half a century, modern cosmology has been chasing shadows of its own imagination. The terms dark matter and dark energy were not born from observation—they were born from *discrepancy*. They emerged as place-holders, standing in for a truth we had yet to comprehend. But the irony is sharp: we mistook misunderstanding for mystery, and built entire empires of physics upon it.

Let's unpack how this illusion was created:

The Logic of Subtraction

Dark matter and energy did not appear in telescopes. They appeared in calculations. When astrophysicists observed the way galaxies rotated, the visible matter didn't explain the speeds at their outer edges. So instead of questioning the assumptions in their gravitational models, they added invisible mass to fix the mismatch.

Likewise, when the light from distant galaxies was found to be redshifting more than expected—implying the universe was accelerating—they invented dark energy to fill the explanatory gap.

This is the logic:

"The result doesn't match our theory → so we add a ghost to the theory."

It's subtractive logic. We define the "dark" not because it's observed, but because something appears *missing.* And that perceived absence is only defined relative to what we think *should* be there.

Echoes, Not Ghosts

Now enter the UED lens. What if the *reason* we see these mismatches isn't due to missing mass or exotic energy—but due to ripple effects we failed to interpret correctly?

- The flat rotation curves are not because invisible matter holds the galaxies together—but because ripple tension persists beyond the visible disk, holding outer stars in synchrony.
- The excess redshift isn't due to some cosmological repulsion—but to ripple fatigue and entropy gradients—slowing and stretching light's alignment over long distances.

In both cases, we are not seeing something missing—we're misreading something present.

Let's use a simple analogy:

Imagine standing in a canyon. You shout. An echo bounces back strangely. You hear two echoes, not one. Instead of concluding the rock formations altered the sound path, you assume a ghost repeated your shout. This is how we created dark matter. This is how we invented dark energy. We heard an echo, and imagined a ghost.

The Consequence of Belief

And so, an entire field of science now chases the invisible—not because it's there, but because the model refuses to let it not be there.

We spend billions trying to photograph a shadow that was only cast because our lamp was tilted. That's the metaphysical trap of conventional cosmology—it holds assumptions sacred, and reality suspect.

As UED insists:

"What you think is missing is simply misaligned. The universe never hid anything. You mistook memory for mass."

It's not a case of the invisible evading us.

It's a case of us building mirrors and then blaming the reflection.

The Metaphysical Trap

Modern cosmology has entered a dangerous philosophical zone—it now defends its models more fiercely than it tests them. Once a field rooted in observation and falsifiability, it has drifted into metaphysics, where unobservable constructs are treated as physical truths simply because the math suggests they *should* be real.

From Science to Safeguarding

At the heart of this trap lies a subtle but damning betrayal: when a theory begins to fail, cosmologists don't discard the model—they add more invisible scaffolding to hold it up.

- Dark matter is unseen? Propose WIMPs or axions—never observed, but mathematically convenient.
- The early universe needs a rapid kickstart? Propose an inflaton field—an abstract scalar field with no traceable source or signature.
- The cosmos expands faster than expected? Insert Λ, the cosmological constant, without a mechanism or cause.

All of these entities have one thing in common:

They are not observed. They are hypothesized to protect what *was already believed.*

This is no longer scientific curiosity—it's intellectual insurance. It's no longer predictive modeling—it's reactive mythology in mathematical clothing.

Mathematical Ontology: A False Equivalence

This trap deepens when math is mistaken for matter.

Λ (the cosmological constant), WIMPs, inflatons—these were once terms in equations, placeholders for forces or effects we didn't yet understand. But over time, these symbols began to gain ontological weight—treated not as unknowns, but as actual entities.

That's the metaphysical error:

To mistake a symbol for a substance. To elevate a variable in a differential equation to the status of physical truth—simply because it makes the math work.

It's the equivalent of writing a story with a fictional character, and then funding an expedition to go find him.

The UED Critique: You're Measuring Echoes, Not Reality

Universal Energy Dynamics (UED) flips this script. It reminds us that you cannot study a structured field by poking its residues.

Conventional tools—colliders, telescopes, detectors—are optimized to sense post-ripple phenomena: light, motion, energy, radiation. But UED claims the real structure lies beneath the echo—in the alignment of the ripple field itself.

You're not going to detect dark matter by smashing protons. You're not going to see dark energy by measuring light shifts. Because neither were *things* to begin with. They were field consequences, not field contents.

To quote the core insight:

"You're not studying the universe—you're interrogating your own instruments."

We keep asking: *"Where is the dark matter?"* UED answers: *"Where did you get the idea that matter must explain everything?"*

We keep looking for missing energy. UED responds: *"You mistook an alignment loss for a cosmic surplus."*

From Empiricism to Alignment Awareness

Real science must return to consequence logic. Instead of looking for particles and forces, we must observe alignment behavior—how energy stabilizes, how entropy gradients shift, how ripple delay explains redshift, motion, and structure.

It's time to admit:

- You can't detect a cause using tools built to detect its aftermath.
- You can't prove reality by chasing ghosts your model required.
- And you can't call it science if you refuse to change the question.

This is the metaphysical trap. UED doesn't walk into it. It steps around it—by asking a better question:

"What is the field trying to realign, and why do we keep missing that message?"

The Observer Bias Problem

In classical science, the observer is assumed to be neutral—merely recording what's "out there" with precision and detachment. But under Universal Energy Dynamics (UED), this assumption collapses. Why? Because every act of observation is ripple-bound.

Observation Is Not Passive—It's Participation

Whether you're looking through a telescope, operating a particle detector, or analyzing sensor data, you're not witnessing a pure, external reality. You're interacting with a ripple field, and your tool is just a local resonance device. Everything you detect is the *effect* of alignment between your instrument and the ripple environment—not an independent truth about the universe.

Let's break this down:

- **Vision**: Photons strike the retina or sensor. But photons in UED are not particles—they're ripple packets, disturbances already shaped by field conditions long before they reach you.
- **Mass detection**: What we call "mass" is just resistance to ripple alignment. You measure inertia or gravity not as absolute forces, but as how hard it is to shift local field tension.
- **Sensors**: Whether it's a CCD in a telescope or an ionization chamber in a dark matter detector, every sensor is built to resonate with ripple gradients. They don't capture SVE itself—they register its imprint.

Key Insight:

"Your instrument doesn't reveal the universe—it reveals what the universe lets through."

We are not neutral observers. We are ripple-bound participants. And we are blind to the stillness, because stillness doesn't resonate. It doesn't emit. It doesn't disturb. It doesn't collide.

Structured Vacuum Energy (SVE)—the still field that underpins all ripples—is undetectable not because it's missing, but because it doesn't need to move to be real.

The Illusion of Empirical Certainty

Scientific empiricism prides itself on being testable, measurable, repeatable. But what we measure is always a filtered echo, shaped by:

- Instrumental bias

- Measurement delay
- Field interaction thresholds

Just as a thermometer cannot measure silence, our detectors cannot see SVE—because they are built on disturbance, not stillness.

UED reframes this clearly:

"Reality is not what you detect. It's what must exist for detection to happen."

The Error of Overconfidence

This leads us to a subtle but fatal flaw in modern cosmology:

- When we can't detect something, we assume it's "dark," "missing," or "non-interacting."
- But UED says: *You built tools to catch fish, and now you're confused why you can't trap the ocean.*

If the structured field is tension-bearing, still, and alignment-based, any ripple tool will only sense after-effects—never the stillness itself.

Hence, looking for dark energy with CMB telescopes or dark matter with cryogenic traps is like:

Trying to measure quiet by amplifying sound.

UED Doesn't Deny Reality—It Redefines What's Reachable

UED isn't dismissing scientific rigor—it's expanding its awareness.

It doesn't say: "Your observations are wrong."

It says: "Your expectations about what's observable were flawed."

UED acknowledges a layered reality:

- Ripple reality: Detectable, measurable, reactive
- Structured reality: Silent, stable, foundational

And the observer sits in between—translating silence into sense, without realizing that their entire epistemology is ripple-dependent.

In conclusion:

"You are not just observing ripples. You are one." And until science recognizes that, it will keep searching for what it was never designed to see.

The Real Experiment Is Realignment

In the age of mega-colliders and space-based observatories, the common assumption is: *If we just build more sensitive instruments, we'll finally detect dark matter or dark energy.* UED says: You're using the wrong language for the question.

Stop Searching. Start Listening to Realignment.

What if the real experiment was never about *detection*? What if it was about inferring what realigned the field—by measuring *consequence*, not *content*?

Universal Energy Dynamics proposes a new experimental protocol: Rather than chasing unobservable particles or stretching theories around unresolvable anomalies, observe the patterns of misalignment, delay, and relaxation.

This new approach rests on three measurable UED parameters:

1. ATT – Alignment Threshold Time

Every ripple structure takes a certain threshold duration to realign with new field conditions. This time lag is not an artifact, but a signature of how deep the misalignment was.

Predicts:

- Delay in field response
- Realignment fatigue
- Dynamic equilibrium zones

Application:

- Planetary drift prediction

- Cosmic filament evolution
- Collapse timings in star clusters

2. RDI – Ripple Delay Index

This index captures phase shift or delay in ripple propagation through tensioned fields.

Predicts:

- Why some light takes "too long" or "too short" to arrive
- Lensing anomalies without visible mass
- Entropic interference zones

Detectable through:

- Redshift phase distortions
- Signal curvature gradients
- Temporal shearing in ripple echoing

3. Entropic Field Topology

This is the ripple terrain map—a structure of where tension has accumulated, diffused, or left memory scars.

Predicts:

- Void behaviors
- Collapse patterns
- Lensing fields with no mass scaffolding

Shift the Scientific Attitude:

"Don't look harder. Look differently."

Conventional science:

- Constructs larger instruments to detect smaller entities.
- Amplifies noise hoping to catch whispers of the dark.

- Thinks more energy or scale = more understanding.

UED's experimental philosophy:

- Study the effect of tension, not the carrier.
- Infer what realigned the ripple field by tracking delay, resistance, and relaxation.
- Abandon force-chasing. Embrace consequence tracing.

From Particle Detection to Pattern Interpretation
You don't need a new telescope. You need a field logic lens.

- Replace Higgs searches with ripple fatigue analysis.
- Replace dark matter halo models with ATT harmonic maps.
- Replace inflation fields with entropic slope reconstructions.

This is not a downgrade of scientific precision—it's an upgrade of orientation.

Closing Remark:

"The most profound shift in science won't be a new discovery. It will be a new way of listening to what was already there."

UED tells us: The real experiment isn't about smashing matter. It's about sensing how alignment breaks, reforms, and carries memory. And that, unlike dark matter, is something we can predict, measure, and apply.

Dark Matter Was a Question, Not an Answer

At its core, dark matter was never a discovery—it was a *placeholder*. A desperate variable introduced not from observation, but from subtraction:

"Something's missing. Let's name it."

From Anomaly to Entity

Let's revisit how dark matter came to dominate cosmological thinking:

- Galaxies were rotating too fast.
- Gravitational lensing was stronger than expected.
- Structures in the universe were too coherent, too early.

Instead of questioning the assumptions behind mass, motion, and space, the mainstream solution was to invent a form of invisible matter to plug the gaps. This placeholder—first a variable, then a hypothesis—quickly solidified into doctrine.

But as UED shows:

The anomalies were never about missing mass. They were about misunderstood field memory.

UED's Bold Position:

Dark matter is not a thing. It was a question, miscast as an answer.

Under Universal Energy Dynamics, what we interpreted as "unseen mass" is re-understood as:

- Ripple tension reservoirs
- Delayed field realignment
- ATT-based dragless harmonics
- Residual alignment zones from past high-entropy events

No need to add a mystery mass—just acknowledge the memory in the field.

Dark Energy: A Relaxation, Not an Explosion

Similarly, what is now labeled "dark energy" was never a measured entity. It was the inverse problem:

"The universe is accelerating. We don't know why. It must be something pushing."

But UED reveals:

- What we're seeing isn't acceleration from a force.
- It's the relaxation of cosmic ripple tension—an unwinding.
- A field de-stressing, not a space stretching.

The ripple field is not inflating—it's calming down. The energy isn't "dark"—it's just done with the noise.

The Failure to Detect Is a Clue

Why, after decades of effort and billions in funding, has no experiment conclusively detected dark matter or dark energy?

Because:

- They are not particles.
- They are not entities.
- They are not *things*.

They are consequences of misreading ripple interactions as mass-related effects.

"If a thing cannot be found, perhaps it was never a thing—just a symptom we misnamed."

Shift the Mindset: From Substance to Consequence

Modern science made a category error:

- It treated effects as causes.
- It tried to see what was meant to be inferred.
- It looked for material where there was only misalignment.

UED corrects that course by showing:

- Lensing doesn't require mass.
- Rotation doesn't require halos.
- Acceleration doesn't require pressure.
- You don't need to invent something to explain your surprise—you need to understand what actually happened.

Final Quote:

"The universe never added invisible matter. It just moved in ways we didn't yet understand."

Closing Insight: Beyond the Detector

"You cannot find silence by amplifying the noise." "You will never detect what isn't hidden—but was never there to begin with."

This chapter closes not in skepticism, but in invitation—an invitation to fundamentally rethink how we seek knowledge from the cosmos.

The End of the Search Culture

Modern cosmology has developed a strange addiction: Search harder. Build bigger. Detect deeper.

Yet for all our effort—miles of underground tunnels, space telescopes the size of buildings, and quantum-level sensors—we're still staring into darkness, asking:

"Where did the rest of the universe go?"

But perhaps the rest was never missing. Perhaps we built an entire scientific culture on the illusion of absence.

We weren't searching for a *thing*. We were searching for a correction—to models that misunderstood reality's architecture.

A New Sensory Paradigm

UED does not claim the universe is unknowable. Quite the opposite—it insists the universe is simpler than we've been taught, but only when observed through the right lens.

This is not a call to mysticism. It is a demand for intelligence over instrumentation.

Instead of brute-force detection, UED advocates:

- Alignment tracing over particle counting.
- Field behavior modeling over entity hunting.
- Consequential logic over subtractive paradoxes.

"Don't ask: Where is dark matter? Ask: What consequence made you think it must be there?"

Final Declaration

The cosmos is not playing hide and seek. There is no missing matter. No hidden force. Just patterns we misunderstood, because we mistook disturbance for substance, and tension for things.

"The universe is not withholding secrets. We are just asking the wrong questions with the wrong tools."

It's time to stop looking through the detector—and start looking beyond it.

Chapter 17: Toward Ripple-Aware Physics

How space missions, quantum research, and cosmology must shift focus to field interaction and ripple response.

Imagine trying to steer a boat by studying the wake it leaves behind.

This is what modern physics does. It builds rockets based on resistance, explains galaxies with shadows, and measures the universe by echoes—insisting that if we just stare hard enough at the turbulence, the ocean's nature will eventually reveal itself.

But the universe isn't turbulent. It's structured. And it doesn't move. It realigns.

From Newton's forces to Einstein's curvatures to the quantum's statistical oddities—we've been circling an answer without ever redefining the question. *What if there's no motion, only memory? No gravity, only gradient? No spacetime, only structured field alignment?*

Welcome to **Ripple-Aware Physics**, where we stop chasing particles and start reading the field.

This chapter is not a criticism of what came before—it is the *continuation they never saw coming.* UED is not here to overthrow science. It's here to **make it listen.**

Let's begin.

A Science of Consequence, Not Constructs

Modern science has made remarkable progress by building constructs—particles, forces, fields, spacetime geometries. But the deeper question remains: Are these constructs fundamental, or are they just useful fictions built upon observable consequences?

Universal Energy Dynamics (UED) answers with conviction: *Physics should be a science of consequence, not constructs.*

Why UED Isn't Fringe—It's a Correction

UED doesn't emerge from speculative metaphysics or exotic equations. It arises from a dissatisfaction with construct-based scaffolding that explains too much by assuming too much. The current standard models in cosmology and quantum theory rely heavily on invented entities—dark matter, dark energy, inflaton fields, Higgs mechanisms—because they try to maintain models after observations diverge.

UED asks a different question: *What if the divergence is real—and your model is what needs correction?*

Ripple logic doesn't patch; it realigns. Instead of correcting symptoms with unseen entities, UED tracks the origin of ripple behavior—why entropy propagates, why field tension relaxes, and how ripple memory drives all apparent forces.

From Newton to Einstein to UED: A Continuum of Refinement

- Newton described force as an external agent causing acceleration—a foundational step, but lacking an understanding of field tension or ripple origins.
- Einstein upgraded this by proposing that space itself bends, and gravity is the result of this curvature—a radical departure, but still built on geometric assumptions rather than structural causes.
- UED now offers the next layer: motion is a perceptual effect of ripple alignment; force is just a resistance to entropy gradients; and gravity is a ripple memory realignment, not curvature.

We don't throw away Newton or Einstein—we realign their insights into a larger field-aware framework.

"You can't rewrite reality, but you can align with it."

— Ripple Principle 1

Scientific Progression Isn't Invention—It's Alignment

True science doesn't invent explanations; it converges toward the truth by peeling off misalignments. Every major shift in physics wasn't about adding complexity—it was about revealing simplicity. Newton unified celestial and terrestrial motion. Einstein unified mass, energy, and spacetime. UED unifies energy, entropy, perception, and alignment through ripple tension fields.

Science must now evolve from constructing models to detecting structural consequences.

Why Consequence-Driven Logic Wins

UED doesn't begin with assumption—it begins with observation of consequence:

- Why does light bend near voids?
- Why do galaxies spin faster at edges?
- Why do redshifts differ without motion?

Mainstream science invents dark scaffolds to explain these. UED observes that all these are consequences of ripple tension—no need for new matter or forces.

In summary, UED represents not a replacement of science, but its re-centering. From inventing theoretical constructs to aligning with structural, measurable consequences, UED invites a return to the core mandate of physics: *understanding what actually is*—not what fits an equation.

The revolution ahead is not conceptual. It's consequential.

Redesigning Space Missions Around ATT

Modern space missions are still grounded in Newtonian kinematics and Einsteinian gravitational curves, even while operating in environments that defy these approximations. Whether it's a Mars slingshot or a lunar descent, the same blind assumptions persist: that space is empty, force is external, and motion can be predicted from mass and geometry.

UED turns this on its head—replacing force calculations with ripple response awareness, and redefining success in space exploration as a matter of field alignment, not brute trajectory.

What Is ATT?

Alignment Threshold Time (ATT) is a UED metric for how long it takes for an object—or ripple disturbance—to reach a new stable alignment within the Structured Vacuum Energy (SVE) field.

It's not about speed or force. It's about how quickly an entity becomes *energetically accepted* by its surrounding ripple topology.

Where classical mechanics looks at acceleration, UED watches realignment latency.

Why Classical Navigation Fails

Classical space missions are designed with:

- Orbital mechanics assuming stable gravitational wells.
- Escape velocity based on simplistic two-body problems.
- Re-entry windows calibrated from drag coefficients and assumed curvature.

This framework ignores ripple memory zones, field gradients, and energetic fatigue present in the ripple field. It's like planning a submarine dive using only GPS altitude—not salinity, pressure, or fluid dynamics.

"In a universe of ripples, thrust is only half the story. The other half is permission."

Real-Time Ripple Mapping > Pre-Set Orbital Prediction

To succeed in deep space or complex maneuvers, missions must map the ripple field live:

- Field tension zones create local resistance or ease.

- Entropy pockets slow or misdirect ripple alignment.
- Ripple fatigue (from past events like starbursts or supernovae) affects field reactivity.

Real-time ripple mapping would allow spacecrafts to:

- Detect zones of low ATT latency (safe re-entry corridors).
- Avoid regions of high ripple resistance (collision risk, drag spike).
- Navigate not by thrust, but by harmonic flow within the field.

Reframing Re-Entry, Slingshots, and Escape Velocity
Under UED:

- Re-entry isn't just drag management—it's a field re-immersion challenge. The craft must descend into a ripple-dense atmosphere without causing phase rupture.
- Gravitational slingshots aren't "gravitational assists" but ripple rebound alignments. The craft enters a ripple zone and exits with its energy redirected, not pulled.
- Escape velocity becomes meaningless—alignment force threshold becomes the metric. You don't escape mass—you exit alignment entrapment.

ATT-Based Control Systems: Avoiding Tragedies
Failures like Chandrayaan-2 or Columbia STS-107 weren't just technical breakdowns. They were misalignments in field understanding:

- Columbia re-entered a turbulent, ripple-fatigued atmospheric zone with no ATT feedback, causing disintegration.

- Chandrayaan-2's lander descended into an unscanned ripple region, experiencing unpredicted resistance during the final phase.

ATT-based control systems could have:

- Measured alignment response delay in real time.
- Delayed or rerouted entry based on ripple calmness.
- Engaged auto-alignment recalibration if phase distortion rose above threshold.

Mission Control in the UED Era

The future of mission control is not thrust calculation—but ripple orchestration.

- Predictive algorithms using RDI (Ripple Delay Index) and ATT fluctuation charts will replace fixed landing windows.
- AI modules will act as realignment assessors, constantly recalibrating the field state, instead of passive inertial guidance.
- Every mission will be a negotiation with the ripple field—not a fight against gravity.

Conclusion:

Space doesn't resist you. It merely reflects how well you align with its memory.

When field behavior becomes the primary input—rather than mass and force—we won't just land on moons or orbit planets. We'll flow into position, like a leaf settling into still water.

"The smartest spacecraft isn't the fastest. It's the one that enters where the ripple field says yes."

Quantum Research Through Ripple Response

The quantum world has long baffled even the most brilliant minds. From wave-particle duality to entanglement, to superposition and tunneling, modern physics has resorted to mathematical formalism and statistical explanations—but not intuitive clarity.

Universal Energy Dynamics (UED) offers a radical reinterpretation:

Quantum phenomena are not "weird"—they are ripple behaviors misunderstood through a particle-based lens.

This section reframes quantum physics as a field of ripple alignments, not object mysteries. Here, quantum events are not probabilistic absurdities, but energetic transitions and memory effects inside a structured vacuum field (SVE).

Why Abandon Wave-Particle Duality?

Wave-particle duality emerged as a patch—a paradox to explain why light behaves like a wave in one experiment (interference) and like a particle in another (photoelectric effect). But duality assumes a binary ontology: something must either be a thing or a vibration.

UED unifies this: Everything is a ripple—sometimes aligned tightly enough to be measurable, sometimes spread thin across field harmonics.

There is no particle, only perceived concentration of field energy due to observer-induced alignment. There is no wave, only a ripple pattern in the still field responding to prior entropic memory.

"You're not observing two natures. You're misinterpreting one ripple."

Quantum Tunneling → Entropy Bleed Zones

In classical physics, a particle cannot cross a potential barrier unless it has enough energy.

In quantum mechanics, it can "tunnel" through. This is treated as a magical probability outcome.

In UED, tunneling is not a leap—it is a bleed.

- The ripple associated with a particle extends into the barrier.
- If the barrier region has entropy fatigue—that is, reduced ripple resistance due to past alignments—the ripple flows through temporarily.
- The result is not motion but re-emergence on the other side, due to field continuity and temporary low ATT threshold.

"Tunneling is not breaking through a wall. It's slipping into a place the wall forgot how to resist."

Quantum Entanglement → Synchronized Ripple Memory

Two entangled particles exhibit instantaneous correlation, regardless of distance. Einstein called this "spooky action at a distance."

Under UED, no spookiness is needed.

- The ripple signatures of entangled pairs share a common origin field.
- Their memory was synchronized when they were created.
- Even if separated spatially, their ATT states remain correlated—responding simultaneously to field perturbations.

There is no faster-than-light communication, only a shared ripple response to changing local alignment conditions.

"Entanglement is not a message sent. It's a memory echoed in two places at once."

Superposition → Temporary Multi-Alignment States

Quantum superposition says a particle can exist in multiple states until observed.

UED reframes this as:

- The ripple hasn't yet resolved into a dominant alignment.

- The system is experiencing simultaneous ripple invitations—but hasn't collapsed into a singular alignment field.
- Observation (or environmental interaction) resolves the tension by choosing the path of lowest ripple resistance.

It's not uncertainty. It's alignment ambiguity.

Superposition is not an object being everywhere, but a ripple distributed across multiple compatible harmonics.

"Superposition is a ripple field asking, 'Where am I most welcome to align next?'"

Predictive Reformulations in UED

Quantum Phenomenon	UED Explanation	Testable Consequence
Tunneling	Ripple continuation through entropy-thinned barrier	Track delay patterns before/after tunneling zone
Entanglement	Shared ripple memory from origin alignment	ATT synchronization measurable in local field shift
Superposition	Multi-alignment invitation within SVE	Collapse response tied to ripple tension map

Summary

Quantum mechanics has long relied on mathematical faith, not mechanistic clarity. But UED restores causality without mysticism.

No magic. No duality. No ghost particles.

Only ripples. Only structured stillness. Only memory, fatigue, and harmonic welcome.

"Quantum reality is not probabilistic. It's patterned response within ripple logic."

In UED, quantum behavior finally speaks a language not of paradox, but of purpose.

Cosmology Without Containers

For centuries, we've told the story of the universe inside invisible boxes—space as a container, time as a river, and matter as passengers riding through both. But these metaphors, however poetic, are not foundational realities—they are conceptual cages.

Universal Energy Dynamics (UED) invites us to escape this prison.

It asserts: There is no container. No flow. No background. Only the structured stillness of the vacuum field, and the ripple interactions that produce meaning.

The Fallacy of the Stretchy Box

In ΛCDM cosmology, the universe expands like a balloon inflating, with galaxies "moving apart" as space itself stretches. This metaphor assumes that space is a thing—a substance—capable of being warped, pulled, and stretched.

But what exactly is stretching? Nothing measurable. Nothing detectable. Just a mathematical placeholder to preserve Newtonian intuition under relativistic terms.

UED proposes: There is no stretching. There is only local ripple realignment—zones of increasing or decreasing tension in a fundamentally still field.

"You're not watching a balloon inflate. You're watching a field remember how to breathe."

Time Is Not a River

Just as space is mistaken for a volume, time is mistaken for a flow—a river carrying events downstream.

But in UED, time is not something that passes. It is how long ripple alignment takes to shift in a local field. In high-entropy zones, realignment is sluggish—so time appears "long." In low-entropy zones, the ripple shifts easily—so time appears "fast."

Hence, "time" is not universal—nor does it "flow." It is a measurement of ripple response delay.

"You're not moving through time. You're syncing with how your field permits change."

Replacing the Big Bang: The Field Ignition Model

Mainstream cosmology starts with a bang—an explosion from nothing, governed by physics that didn't yet exist.

UED sees this as an epistemological failure: You can't explain a field's beginning from a state that lacks alignment logic.

Instead, UED posits a more elegant principle:

- The still structured vacuum (SVE) was always present.
- What we call the "beginning" was not a creation, but the first major alignment event—a ripple ignition that began modulating tension across the field.

The universe didn't "explode into being." It entered resonance.

And like all ripple systems, its early zones bore high tension, which gradually relaxed—giving rise to the illusion of expansion and directional time.

Cosmic Evolution = Tension Modulation

In UED:

- Cosmic structures emerge where tension aligns harmonically.
- Galaxies, voids, and filaments are regions of structured ripple activity, not "objects in space."
- Evolution happens when the field relaxes or sharpens its internal tensions.

So instead of spacetime expanding, we observe:

- Local entropy modulation

- Ripple field breathing patterns
- Realignment sequences across harmonic thresholds

And this changes everything—from how we model cosmic microwave background radiation to how we interpret galactic redshift.

"The cosmos didn't grow. It remembered its tensions—and resolved them, slowly."

The Paradigm Shift

Old Paradigm	UED Paradigm
Space is a fabric	Space is a tension topology
Time flows	Time measures ripple realignment latency
Big Bang	Initial ripple ignition in still field
Expansion	Field relaxation from high-entropy pulse
Movement through space	Reconfiguration within field states

This isn't just a semantic upgrade. It's a seismic rethinking of existence itself.

UED reframes cosmology not as the study of things in space, but as the study of how field tensions birth form, memory, and alignment.

"The universe didn't begin. It began aligning."

That distinction marks the boundary between object thinking and field thinking—and it's the bridge to a ripple-aware future.

Engineering Ripple-Responsive Instruments

The age of particle obsession is ending.

For over a century, physics has hunted for the building blocks of the universe—quarks, bosons, WIMPs, axions. We built *detectors*, not *understanders*. We watched for *collisions*, not *consequences*. Our machines see what's thrown at them, not what breathes through them.

But UED (Universal Energy Dynamics) brings a different engineering imperative:

"The universe is not made of things. It's made of alignments."

So we must build tools not to detect objects, but to sense alignment changes—to capture the consequences of ripple tension, not just its energetic spikes.

Let's outline the foundation of Ripple-Responsive Instrumentation.

Detectors of the Future: Field-Aware, Not Particle-Sensitive

Legacy detectors—like LHC sensors, cryogenic dark matter chambers, or CMB telescopes—are entirely interaction-bound. They require something to *hit*, *emit*, or *radiate*.

UED reframes the question.

Instead of asking, "What passed through the detector?" we ask:

"What changed the tension pattern of the field?"

This shift moves us away from event detection toward field memory analysis.

The tools must evolve.

Key Components of Ripple-Responsive Engineering

1. **RDI Sensors (Ripple Delay Index):** These sensors measure the latency of ripple propagation across structured vacuum zones. Rather than tracking mass, they map field responsiveness. High RDI → fatigued, compressed, or recently disturbed zones. Low RDI → relaxed, pre-aligned, or high-permeability regions.
2. **ATT Loggers (Alignment Threshold Time):** Every object realigns only when local ripple tension reaches a specific threshold. ATT loggers don't record energy—they record when alignment shifts become possible. They're the new chronometers of field behavior.
3. **Entropy Topology Mappers:** Inspired by geological

surveys and fluid flow models, these instruments map the energy fatigue and gradient zones of SVE. The goal is to visualize ripple congestion, predict realignment fatigue, and detect phantom influences without invoking mass.

Together, these tools could replace mass reconstructions, fake halos, and statistical guesswork with live tension topography.

Prototype Suggestion: The Ripple Alignment Observatory

Imagine a new kind of observatory—not looking for light, radiation, or noise...

...but scanning the cosmos for ripple harmony and fatigue.

The Ripple Alignment Observatory (RAO) would:

- Map ATT and RDI across galactic fields in real-time.
- Track cosmic events as tension reconfiguration, not mass transfer.
- Identify entropic scars left by past collapses or mergers.
- Forecast high-instability zones where black hole jets, flares, or warps are likely—not by watching light, but by reading ripple unease.

It could also serve Earth-based uses:

- Monitor geomagnetic field disruptions as ripple anomalies.
- Track seismic pre-patterns via subsurface tension misalignments.
- Predict satellite drag shifts and communication dropouts by watching field turbulence, not just atmospheric density.

"The field remembers what your instruments forgot to ask."

Implication: From Looking to Listening

Ripple-responsive instruments won't observe in the classical sense. They will listen—to the latent harmonics of the field. To the memory of motion, not the motion itself.

We are engineering not sensors, but synchronizers.

Tools that don't wait for a collision or signal. Tools that sense the readiness of the field to express a shift.

"Stop looking for particles. Start sensing the permission of reality to change."

This is the engineering future under UED. Where we finally stop trying to detect ripples as noise—and begin understanding their silence as signal.

The Ethical and Epistemic Shift

Universal Energy Dynamics (UED) doesn't just revise physics—it reframes how we should do science.

We've long inherited a worldview where science is conquest:

- To *master* nature.
- To *extract* truth.
- To *dominate* uncertainty.

But UED exposes a deeper truth:

We never studied nature—we disturbed it into appearing.

This section explores the ethical and epistemological transformation required for ripple-aware science.

From Discovery to Alignment

Conventional science operates under detached empiricism:

- "We are neutral observers."
- "Reality exists independently of us."
- "Data is passive. We're just measuring."

UED obliterates that assumption.

In ripple physics, there is no neutral observer. Every act of observation—whether collision, detection, or measurement—is a ripple intrusion, a realignment attempt that echoes through the field.

"You don't observe the ripple. You become part of its next configuration."

Thus, science must shift from:

- Discovery → Discernment
- Intervention → Intuition
- Force → Field Sensitivity

Observer = Participant = Ripple Influencer

Every instrument you build...Every question you ask...Every assumption you project into the model...

—alters the very system you claim to observe.

This is not metaphysics. It's epistemic realism under UED:

- Ripple delay changes with tension.
- Entropy propagation is observer-sensitive.
- Measurement induces misalignment unless designed to harmonize.

The scientist is not outside the system. They are the initiator of its next wave.

Hence the ethical responsibility: To observe without disturbance. To align, not impose.

Building Knowledge Without Distortion

We've been distorting reality by:

- Reducing it to force vectors.
- Treating mathematical models as ontological truth.
- Searching for evidence within the consequences we create.

UED demands a new ethos:

- Don't model what you can't feel in field response.
- Don't simulate what you haven't synchronized with.
- Don't amplify noise and call it signal.

It's time to stop excavating truth with force—and start receiving it through sensitivity.

Education Reform: From Formulas to Field-Sense

Current education, especially in physics and cosmology, teaches:

- Equations as sacred.
- Theories as absolute.
- Observation as detached fact.

UED challenges that pedagogy.

Instead, we must teach:

- Ripple logic over rigid math.
- Field tension intuition before measurement.
- Consequential reasoning instead of construct memorization.

Students must learn to feel alignment shifts in real-world systems:

- Watch a pendulum and see ripple tension recovery—not just simple harmonic motion.
- See thermodynamics as entropy memory—not statistical chaos.
- Understand electricity not as charge flow—but as ripple displacement and field response.

"We don't need better formulas. We need better feelers."

A New Scientific Humility

What if our greatest discoveries were just tolerable distortions of the field? What if our most advanced tools merely detect echoes, while the real reality remains unshaken?

UED offers not despair—but dignity.

It shows that true understanding doesn't come from power—it comes from participation.

To be in science now means:

- Accepting your ripple effect.
- Calibrating your epistemic noise.
- Dedicating yourself to ethical alignment, not conceptual conquest.

Closing Reflection: Science as Subtle Listening

We end with a final provocation:

"We are not just discovering reality—we are disturbing it less accurately."

And the more we reduce our distortion, the more reality begins to whisper back.

UED invites science into its next maturity—not as an industry of knowledge, but as a discipline of resonance.

Where to know is not to seize... but to sync.

Call to the Global Scientific Community

This is not just another theory. It is a fundamental reframing of how science sees the universe.

Universal Energy Dynamics (UED) doesn't compete with current models—it reveals the underlying pattern they all attempt to approximate. It speaks not in symbols, but in consequences. Not in particles, but in patterns of alignment. It invites a recalibration of the scientific endeavor itself.

And now, the time has come to invite collective participation.

Rewrite Theoretical Models Using UED Principles

We urge theoretical physicists, cosmologists, quantum researchers, and applied scientists to begin rewriting existing frameworks—across domains—under the logic of ripple alignment.

This means:

- Replacing invisible entities (dark matter, inflatons, gravitons) with measurable ripple delays and entropy gradients.
- Reformulating field equations based not on force interactions but on alignment thresholds (ATT), Ripple Delay Index (RDI), and tension topography.
- Recasting time, mass, and motion as field memory, resistance to alignment, and localized ripple fatigue.

"Science must return to consequence—not assumption. To tension, not object. To structure, not scaffold."

No more fudge factors. No more exotic placeholders. Let ripple logic clear the fog of forced explanations.

Collaborate on Open-Source Ripple Map Projects

The next era of scientific instrumentation won't just capture radiation. It will trace ripple flows across space and time.

We propose the formation of an open-source global consortium to develop:

- Ripple Mapping Software – integrating RDI, ATT, and entropy field fluctuations.
- Field Response Observatories – designed to detect and log phase delay zones, entropic shifts, and alignment memory signatures across cosmic and terrestrial scales.
- Realignment Simulators – tools that visualize how ripple fields evolve, not how objects move.

This is not just theoretical. This is an engineering roadmap for ripple-aware technology, satellite navigation, atmospheric modeling, and material sciences.

"Every scientific discipline is ripple-sensitive. They just don't know it yet."

Let's build the maps reality has always been waiting for.

Invite Journals, Institutions, and Policy Bodies to Reframe

We call upon:

- Scientific journals to create open sections for ripple-based and consequence-aligned research.
- Academic institutions to fund interdisciplinary programs on structured vacuum theory, ripple mechanics, and consequence-based metaphysics.
- Space agencies and policy bodies to shift research funding from "detection" of unprovable phenomena toward the development of ripple-traceable models and missions.

Science isn't just about what we fund. It's about what we dare to reimagine.

The cost of continuing to chase dark ghosts is not just financial. It's intellectual, moral, and generational.

A Gentle Closing Provocation

This is not a rebellion.

It's not about breaking away.

It's a return—to the logic the field always followed. Before equations tried to mimic it. Before institutions boxed it. Before cosmology distorted it.

"UED is not a revolution. It's the field finally being heard in its own language."

And that language is simple:

- Realign.

- Respond.
- Remain true to consequence.

We now hand this torch to the global scientific community. Not to convert—but to co-create. Not to debate—but to observe with new eyes.

The invitation stands: Come ripple with us.

Closing Reflection: From Motion to Memory

"In ripple-aware physics, we stop chasing particles and start tracing patterns." "The universe isn't expanding or shrinking. It's remembering—and we are finally listening."

In the end, what we thought was motion was memory all along.

Motion—a cornerstone of classical physics—was never about things *going somewhere*. It was always about *something aligning differently*, creating the illusion of travel, force, or displacement.

UED gently removes the illusion without invalidating the utility. It doesn't say, "Motion never existed." It says, "What you called motion was ripple realignment—field memory adjusting to entropy."

The Death of Force, The Birth of Form

What dies with UED is not science—but the myth of object-based causality.

- Forces become shorthand for response to tension differentials.
- Particles become local condensation of aligned ripple information.
- Time becomes a trace of transitions, not a flowing river.
- Space becomes the map of remembered distortions, not an expanding void.

We stop measuring how fast something goes, and start asking, "How long did the field take to let that happen?"

Listening to the Field's Memory

- **Cosmic expansion?** Perhaps it's not a stretching fabric—but a relaxing field, breathing out the tensions of ancient ripple conflicts.
- **Quantum randomness?** Maybe it's localized ripple fatigue, unable to sustain a single alignment.
- **Dark energy?** Try entropic memory release—the field remembering how to become still again.

A New Way to Be Scientific

In ripple-aware science:

- We don't build bigger colliders—we build sharper attention to ripple transitions.
- We don't worship constants—we explore contextual response thresholds.
- We don't cling to predictions—we stay humble before consequences.

We begin to understand that the universe is a memory foam of field interactions. What you do to it stays imprinted—until it realigns.

And the job of science is no longer to extract truth from phenomena, but to observe without disturbing, to trace patterns without inserting noise.

Final Realignment

Let this be the moment where science paused—not to surrender, but to listen.

To ask not, "What is it made of?" But, "What is the field still remembering here?"

To stop seeking new particles and start noticing old alignments.

Because maybe the universe isn't growing or decaying.

Maybe it's simply finishing a thought. And with UED, we finally learned how to read its handwriting.

And that... is the final shift: From motion... to memory. From explanation... to alignment. From chasing answers...to becoming the stillness that already holds them.

Chapter 18: You Were Looking in the Wrong Direction

Final summary of how "dark" is not hidden, but simply a signpost toward the limits of classical observation.

We have spent the better part of a century chasing shadows—assigning mass to silence and motion to uncertainty. Telescopes have peered into the void, detectors have tuned themselves to whispers that never arrive, and physicists have stretched equations to account for what refuses to be seen. But what if the failure to find was never a failure of technology—but of direction?

Dark matter was never missing. Dark energy was never hiding. They were never there to begin with—not as entities, but as signs that something else was misread.

We were looking in the wrong direction—not spatially, but conceptually.

Not toward the field, but away from it. Not toward structure, but toward substance. Not toward consequence, but toward cause.

This chapter is not a conclusion. It is an orientation reset. A call to stop gazing deeper into the same tunnel, and instead to turn—slowly, silently—toward the ripple patterns left all around us.

Because darkness was not a wall. It was a message. And now, we read it differently.

What Was "Dark" Ever About?

The word *"dark"*—in "dark matter" and "dark energy"—has captivated scientists and the public alike, evoking images of a hidden universe, mysterious forces lurking just beyond our observational reach. But in truth, *dark* was never about literal darkness. It was not about the absence of light. It was about the absence of explanation.

Let us strip away the myth and return to the origin. "Dark" was a placeholder. It entered the scientific lexicon not as a discovery, but as a confession. A quiet nod to everything the models couldn't explain.

When galaxy rotation curves remained flat instead of declining, when distant supernovae appeared dimmer than expected, when gravitational lensing appeared in regions devoid of mass—science did not adjust its assumptions. It did not question its framework. It baptized the anomaly. *Dark matter. Dark energy.*

But these names were never revelations. They were apologies dressed as discoveries.

"Dark" became a conceptual bandage over mathematical bruises. Not because the data was wrong, but because the lens through which it was interpreted was flawed. These anomalies weren't signs of exotic hidden forces. They were fingerprints of misunderstanding—visible consequences of ripple field mechanics that classical physics couldn't read.

Imagine a detective solving a crime not by reexamining the crime scene, but by inventing an invisible suspect to match the outcome. That's what "dark" has done to cosmology.

The tragedy is not that we didn't see something. It's that we saw something *clearly*—and misunderstood it. The rotation curves were real. The lensing was real. The cosmic acceleration was observable. But the attribution was incorrect. Not because of malice or incompetence, but because the tools and frameworks were ripple-blind. They could only interpret reality in terms of motion, mass, force, and flow—not alignment, tension, or ripple memory.

"Dark" was not darkness. It was diagnostic. A symptom. It revealed where the paradigm cracked.

And in that light, we must realize: what was called dark, was never hiding. It was *pointing*. Pointing to the edges of Newton. The edges of Einstein. The edges of our epistemic comfort.

The moment we treat "dark" not as a substance to be found, but as a boundary condition of classical error, the illusion dissolves. We weren't peering into a dark universe. We were staring at the limits of our own language—and calling the silence strange.

Limits of Observation, Not Reality

The core flaw in our scientific pursuit of the "dark" lies not in the data—but in the very *architecture* of our detection systems. Classical science, from its Newtonian roots to its quantum edge, is built entirely on interaction. If something cannot nudge, disturb, emit, reflect, or resist, it is declared non-existent.

Our instruments, therefore, were designed with a bias: they are fundamentally ripple-blind. They detect only the aftermath—the echoes, the effects. Lightwaves, force exchanges, thermal emissions, gravitational disturbances. But what if the truth precedes all of that? What if the origin of reality—what UED calls Structured Vacuum Energy (SVE)—doesn't announce itself with a bang, but remains perfectly silent?

Classical tools demand interaction to confirm existence. But the ripple field operates through *realignment*, not interaction. A zone of perfect equilibrium—a region of no misalignment, no ripple fatigue, no entropy gradient—will appear utterly invisible to such tools. Not because it isn't there, but because it has no reason to echo back.

This is the tragic irony of conventional observation: we built flashlights to study silence. We used vibration detectors to measure stillness. And when the silence didn't respond, we concluded it was missing.

But UED offers a radical reframe: the *failure* to detect is not a failure of reality—it's a failure of method. And more provocatively, it's actually a *signal*—not of what exists, but of *what doesn't need to participate.*

Let's reimagine "darkness" not as something lurking in the shadows, but as a calm pond in a world trained to recognize only splashes. If your tool cannot detect a ripple, perhaps it's because the field is at rest. Perhaps darkness is equilibrium.

In UED terms, "dark" regions often represent non-participating field zones—areas in perfect alignment, or post-ripple recovery, where energy no longer fluctuates visibly. These zones do not emit or resist. They don't act—they *are.*

Here, failure becomes diagnostic. If your gravitational lensing data yields no mass, but a visible bend, maybe the field itself retained a memory of prior tension. If your detectors find no particles, perhaps they're scanning a zone of absolute harmony.

Structured Vacuum Energy does not offer itself to brute detection. It invites interpretation through field behavior, ripple tension, and the subtle choreography of delay and alignment.

To put it plainly: we were not observing the universe. We were interrogating only the noisy parts. We called what was undisturbed "missing." And in doing so, we mistook the edge of our reach for the edge of reality.

When We Mistake Gaps for Ghosts

Let's confront the core mistake—one that modern cosmology repeats like a nervous tick: whenever there's a mismatch between prediction and observation, it doesn't question the prediction. It fabricates a ghost.

Gravitational lensing shows up without enough visible mass? Invent "dark matter." The universe's redshift doesn't align with simple expansion models? Add "dark energy." Galaxies rotate too flatly for their mass? Insert "halos" of invisible substance.

But what if none of these effects required *things*? What if they were merely misunderstood *responses*?

Let's decode this one by one.

- **Lensing does not imply mass.** It implies *ripple distortion*. When light bends, it's not curving around a mass—it's flowing through a ripple-tension gradient. Like wind adjusting around a pressure zone, not due to gravity, but due to ripple memory.
- **Redshift does not imply recession speed.** It implies *ripple fatigue*. Light stretches because it travels through tired fields—zones where energy is still relaxing from past disturbance. You're not watching a galaxy speed away; you're watching a message crawl through molasses.
- **Gravitational pull does not imply force.** It implies *field memory*. What we perceive as gravity is simply the tendency of the ripple field to restore alignment. It's not attraction. It's recollection.

But here's the twist: mainstream science never asked, *"What field effect could be causing this?"* Instead, it asked, *"What particle must exist to account for this?"*

That's the **inversion error**—mistaking a symptom for a substance, a ripple for a rock, a gap in comprehension for an undiscovered ghost.

In UED, these so-called dark phenomena are not primary realities. They are *derivatives*—expressions of ripple field dynamics misunderstood through the narrow lens of force-based, object-centric paradigms.

This misalignment led to a deeper trap: a metaphysical loop where failure to observe something became proof that it *must* exist somewhere deeper, fainter, or more elusive. We began looking *deeper* instead of *wider*. We stared harder into the tunnel, never questioning if the tunnel itself was real.

Imagine mistaking a shadow for a person. Then imagine funding a century-long expedition to study the "species" of that shadow—building sensors to detect its heartbeat, crafting simulations of its lifestyle, and naming it "dark sapiens." All because the light hit an angle you didn't understand.

That's where we are. And that's what UED corrects.

It doesn't hunt the ghost. It removes the fog.

It sees lensing, redshift, and gravity not as mysteries requiring more mass, but as ripple expressions requiring better models.

You don't need to invent particles for effects born from field behavior. You need to realign your thinking to the medium itself.

Dark matter was never a substance. It was a question... asked backward.

Reframing the Scientific Horizon

Every failed detection should have been a clue. Every null result from the world's most sensitive dark matter detectors. Every contradiction emerging from increasingly desperate dark energy models. Every "error bar" stretched beyond comfort—none of it was failure.

It was the universe whispering: "Your map ends here—not because I do, but because your method does."

Mainstream science, trained in mechanical metaphors, saw the cosmos as a machine. A grand, gear-filled clockwork of forces, particles, velocities, and predictable equations. But machines obey logic only within the blueprint they're built on.

The universe never had a blueprint. It had a memory field—a ripple-responsive, self-balancing tension network that doesn't run like a machine. It *remembers*. It *adjusts*. It doesn't move. It *realigns*.

UED invites a radical shift—not just of theory, but of *attitude*.

- From discovering to listening. Stop assuming the universe is hiding answers. Start accepting that our models might be making noise where there was only stillness. Discovery in ripple-aware physics is not a conquest—it is a *humbling alignment*.

- From detecting to aligning. The obsession with detection—particles, signals, high-energy impacts—comes from an outdated view that reality must hit us hard to be real. But UED shows: *reality flows in fields*, not collisions. Detection now means tuning into ripple gradients, entropy shifts, and ATT/RDI harmonics—not triggering an event in a chamber.

This shift reframes the scientific horizon.

- The end of what we could "see" wasn't the end of what's there.
- The failure to isolate dark particles wasn't an absence—it was *resistance* to being understood with inappropriate tools.
- The mathematical fudge factors weren't gaps in nature. They were *patches on broken frameworks*.

What mainstream science calls the "unknown" is often just the uninvited—phenomena excluded by design, filtered out by assumptions, or ignored because they don't shout in photonic language.

UED proposes a different ethic of science:

- One that stops demanding that the universe behave like a Newtonian lab.
- One that listens for patterns, instead of forcing data to fit equations.
- One that accepts that alignment is not an effect—but a condition of understanding.

Let's be clear: this is not mysticism. It's epistemological hygiene.

The tools weren't wrong. They were aimed in the wrong direction. They were built for noise, not nuance.

The classical horizon said: "We can see up to here." UED responds: "That's where your flashlights stop—not where the field ends."

When we treat the cosmos as a memory field instead of a mechanical stage, the so-called edge of the known becomes not a wall—but a mirror.

And in that mirror, we begin to see: Darkness wasn't the unknown. It was our method, misaligned.

Darkness as a Teacher

For centuries, darkness was treated as a threat.

In ancient mythologies, darkness symbolized fear, ignorance, or evil. In modern science, it became a placeholder for ignorance—a sign that our models didn't work. So we declared war on it. We built machines to "illuminate" the dark, to penetrate it, to capture its secrets.

But what if darkness was never the enemy? What if it was... the instructor?

What if darkness—the dark matter, the dark energy, the unseeable voids, the unexplained gaps—was *never meant to be solved*, but meant to make us pause?

"You were not looking in the wrong place. You were using the wrong kind of light."

That single shift—from predatory detection to reflective perception—is where UED begins to reshape our understanding.

Respecting the Dark

Imagine respecting darkness instead of fearing it. Treating the unknown not as a failure of science, but as a feedback mechanism. Something that doesn't withhold the truth—but invites us to confront our limitations.

In that reframing:

- Darkness becomes a teacher of humility.
- It reminds us that not all knowledge is visible.
- It whispers that the absence of interaction is still information.

Three Lessons from the Dark

UED doesn't just ask new scientific questions. It introduces a new posture. One that values three radical lessons darkness teaches us:

1. Not all things are "seen." The ripple field—Structured Vacuum Energy (SVE)—does not emit, reflect, or resist in classical ways. It is *tension without motion*. Its reality is not observable—it is consequential. If it exists, you don't *see* it. You *experience its harmony or its dissonance*.
2. Absence of effect does not mean absence of cause. A particle detector that finds nothing is not reporting emptiness. It's revealing alignment—a lack of disruption in the field. Classical science saw this as failure. UED reframes it as *successfully reaching stillness*.
3. Sometimes, stillness is the result. The culmination of all interactions is not more motion. It is relaxation. When a ripple settles, there is no noise, no light, no resistance. That's not a dead zone. That's perfection—the field, returned to balance.

Humility Over Hubris

UED's greatest offering is not just technical. It's philosophical. It doesn't seek to dominate the unknown, but to realign with it.

It says:

- We are not the masters of the cosmos.
- We are participants in a ripple memory far older and subtler than detection can trace.
- Our job is not to "explain away" the dark.
- Our task is to ask better questions, with quieter tools.

So when future scientists face a mystery—an unexplained redshift, a lensing arc with no mass, a galaxy too mature to exist—they should no longer build a ghost to fill the gap.

They should ask:

- What if this is a ripple memory?
- What if the field aligned, and I'm just late to feel it?
- What if this darkness... is truth in stillness?

We thought we were exploring the dark. But it turns out: The dark was exploring us.

The Realignment of Curiosity

For centuries, curiosity has driven humanity to build bigger, sharper, louder tools in search of universal truths. We launched billion-dollar telescopes to see farther. We built underground detectors to listen harder. We collided particles at near-light speeds just to crack open the smallest mysteries.

But what if we've mistaken volume for insight? What if scientific revolutions aren't born from bigger equipment—but from better alignment?

"The real frontier isn't out there. It's in the *questions we stopped asking*."

Universal Energy Dynamics (UED) doesn't expand the search. It redirects the searchlight inward—toward the assumptions that shaped the search to begin with.

Science Needs a Realignment

We say we want to know the universe. But how do we ask?

Here are three ripple-aware questions that can replace entire chapters of classical confusion:

1. What if space is not a thing—but a ripple memory?

Classical physics treats space as a canvas: A neutral backdrop stretched, curved, or torn by matter and energy.

But UED repositions space not as a thing, but as the residual tension of ripple interactions. It's a memory of past alignment, not a passive void.

So instead of stretching space to explain anomalies, UED asks:

- What *field configurations* existed before?
- How long has this zone been at rest or under tension?
- What ripple memory are we standing inside?

This reframes space not as *geometry*, but as topological memory.

2. What if energy is not applied—but released through misalignment?

In Newtonian logic, energy is something added: Push harder → more energy.

But in UED, energy is revealed, not injected. Every ripple, every particle, every emission is a field's attempt to relax from stress.

In that view:

- Energy is not a force—but a consequence of misalignment.
- A supernova is not a power surge—but a field scream to release ripple tension.
- Even a simple photon isn't "sent"—it's *spilled* from imbalance.

This reframes energy as a symptom of disequilibrium—not a magical currency.

3. What if time is not flowing—but tracking the pace of relaxation?

What is time, really? Not a ticking clock. Not a fourth dimension.

UED suggests: "Time is the measure of how quickly tension realigns."

Where there's high entropy—time "moves fast." Where there's equilibrium—time stands still.

It's not the universe evolving *through* time. It's the universe relaxing, and time is just the ripple wake we use to track it.

Thus:

- Time dilation is ripple distortion, not spacetime bending.
- Early galaxies don't "defy time"—they reflect zones of early alignment.
- Black holes don't "stop time"—they compress ripple memory beyond update.

A New Vocabulary for a Silent Field

When we used the word *dark,* we really meant: "I don't know what this is, so I'll call it something dramatic."

But UED invites a realignment of curiosity—and that starts with new language:

Instead of:

- "Dark matter" → say: "Ripple memory field not yet misaligned"
- "Dark energy" → say: "Entropy-driven relaxation"
- "Expansion" → say: "Alignment redistribution"

You don't need dark in a ripple-aware universe. Because in a still field, illumination is optional.

What you called *dark* was simply what the field hadn't needed to light.

The mind doesn't need brighter torches. It needs cleaner mirrors.

We were not stupid to ask the questions we did. But we must now be brave enough to ask the ones we didn't.

Because the universe was never hiding anything.

It was waiting for us to ask better.

Closing Reflection: You Were Never in the Dark

"You didn't fail to find it. You failed to ask what you were really seeing."

For decades, perhaps centuries, our greatest scientific minds stood beneath the cosmic sky—puzzled, persistent, and profoundly committed. We saw galaxies spinning too fast. We watched light bend where there was no visible mass. We measured the cosmos and found a strange, accelerating drift.

And when nothing in our known equations could explain it...We declared: "Something dark must be out there."

But darkness was never the problem.

A Universe Never Meant to Be Lit

What if the universe never intended to shine a spotlight on everything? What if the quiet spaces, the still zones, the equilibrium states—were not hiding—but simply not *trying* to be seen?

We thought the cosmos was withholding something. But it was only reflecting something back:

Our *own limitation*—Our obsession with force, light, movement...Our fear of silence. Our discomfort with memory, with stillness, with alignment we couldn't measure.

What We Chased Was Already Around Us

We didn't miss the dark matter. We didn't miss the dark energy.

What we missed was the message:

That our methods had peaked—not our knowledge.

We had mistaken *what resists detection* for *what must exist.* We thought absence of proof meant presence of mystery. We confused failure of instruments for failure of nature.

And so, we built more instruments. Bigger colliders. Sharper telescopes. Deeper detectors.

But we were always searching *through the wrong lens.* We were tuning our senses for flashes, while the truth waited in frequencies of stillness.

The Mirror, Not the Veil

Let's be honest.

The "darkness" was never a veil the universe drew to taunt us. It was a mirror—held gently and patiently—asking:

"Will you finally look at yourself, before you accuse me of hiding?"

Dark energy was not a substance. It was a placeholder for misalignment. Dark matter was not a particle. It was a question dressed up as an answer.

We weren't in the dark. We were looking for light in a place that *wasn't built to glow.*

Final Truth: The Light Wasn't Missing—Your Gaze Was Misaligned

The cosmos never hid anything from you. You were just asking the wrong questions. With the wrong assumptions. Using tools designed for disturbance instead of alignment.

You trained your eyes for sparks. But forgot that reality—deep, universal reality—is shaped in silence.

Stillness was never empty. It was full—of meaning, memory, and consequence.

And now, with UED, we don't just turn on a new light. We *turn ourselves*—in a new direction.

Not to chase what we couldn't see. But to finally feel what was always shaping us.

The field. The ripple. The gentle, unlit thread of everything.
So let's say it clearly, at last:
You were never in the dark. You were just facing the wrong way.

Epilogue: The Field Was Always Speaking

You came to this journey expecting darkness. Expecting revelations of unseen particles. Expecting grand theories to fall, or new constants to rise.

Instead, you found something simpler. Not a theory, but a correction. Not a revolution, but a realignment.

Science, in its hunger, had made the universe a problem to solve. But the universe was never a puzzle. It was a pattern.

A ripple. Not shouting. Not hiding. Just *being*.

For centuries, we declared that motion defined existence. We chased orbits, spins, speeds, explosions. We measured everything that moved—and ignored what held it all together.

But stillness was not emptiness. Silence was not absence. What we dismissed as "dark" was simply what hadn't flared up for our tools.

UED didn't tell us there is no motion. It told us motion is the result, not the reason. That mass doesn't pull. That energy doesn't push. That space doesn't bend.

What changes—is alignment. What persists—is memory. What connects—is the field.

Now, we know better.

The cosmos is not expanding. It is exhaling entropy. The galaxies are not drifting. They are echoes of earlier harmonies. And light is not a messenger—it is a ripple trail, left behind in a responsive medium.

You Stand at a New Threshold

This was not a book. It was a turning of gaze.

From particles... to patterns. From force... to field. From discovery... to discernment. And perhaps most importantly—from observation to participation.

You are no longer just a spectator of the universe. You are a ripple within it. A conscious phase alignment. A frequency through which memory becomes motion.

Final Pulse:

The universe never needed dark matter. It never needed dark energy.

It only needed you to stop naming what you feared and start listening to what was always whispering through the silence:

"I never moved. You just started noticing."

APPENDICES

Appendix A: Ripple Equations Replacing Λ and G

New formulations based on ripple gradients, Alignment Threshold Time (ATT), and entropy divergence.

Purpose

To introduce alternative, consequence-rooted equations that replace the metaphysical constants Λ (cosmological constant) and G (gravitational constant) with UED-native variables: ripple gradients, field alignment metrics, and entropic topology. These equations aim to offer predictive, testable models rooted in real-time energy field behavior—rather than static universal parameters.

I. Rethinking Constants

Why Λ and G Fall Short:

- **Λ (Cosmological Constant)**: Introduced as a fudge factor to account for accelerating cosmic expansion—yet its source remains unmeasurable and its units ill-defined in physical terms.
- **G (Gravitational Constant)**: Treated as a universal "glue" constant, but it is not derivable from any field principle—its value must be inserted empirically.

II. Ripple Field Equation 1 — Local Field Curvature by Ripple Gradient

UED Equivalent to Einstein Field Equation:

Instead of:

$$G_{\mu\nu} + \Lambda g_{\mu\nu} = \frac{8\pi G}{c^4} T_{\mu\nu}$$

UED proposes:

$$\mathcal{F}_{\mu\nu} = \kappa \cdot \nabla_\mu \mathcal{R}_\nu - \eta \cdot \partial_\mu \partial_\nu \sigma$$

Where:

- $F_{\mu\nu}$: Ripple tension tensor (field response)
- R_ν: Local ripple memory vector
- σ: Entropy gradient scalar field
- κ: Constants of ripple responsiveness calibrated to ATT and RDI

Interpretation: Realignment (instead of curvature) dictates energy dynamics. Spacetime is not bent—it's tensioned. Mass-energy acts as a ripple anchor, not a warper.

III. Ripple Field Equation 2 — Gravitational Effects as Ripple Memory Drift

Instead of Newton's:

$$F = G \cdot \frac{m_1 m_2}{r^2}$$

UED reframes as:

$$F_{\text{eff}} = \delta E \cdot \left(\frac{1}{R^2} \cdot \Delta \text{ATT} \right)$$

Where:

- δE: Energy mismatch between two field zones (not mass)
- R: Ripple path curvature radius
- ΔATT: Difference in Alignment Threshold Time (field responsiveness)

Interpretation: What we call gravitational force is actually the resistance to alignment shift between two ripple-anchored systems.

IV. Replacing Λ — Field Relaxation and Entropy Divergence

Classical cosmology claims accelerating expansion due to Λ. UED claims:

$$\text{Apparent acceleration} = \frac{d\mathcal{A}}{dt} \propto -\nabla\sigma$$

Where:

- A: Alignment density (field coherence)
- $\nabla\sigma$: Entropy divergence (field relaxation tension)

Interpretation: As the field entropy relaxes, objects appear to move apart—not because they are pushed, but because their shared ripple memory is weakening.

V. Realignment Energy Formula (UED Work-Energy Equivalent)

$$\mathcal{W}_{\text{UED}} = \int \delta\mathcal{T}(x) \cdot \frac{d\mathcal{R}(x)}{dt} dx$$

Where:

- $\delta T(x)$: Local tension differential
- $R(x)$: Ripple memory state
- dR/dt: Rate of re-alignment

***Interpretation*:** Work is not mass-based force over distance, but ripple tension realignment over entropy shift.

VI. Summary Chart — Ripple Variables vs Classical Constants

Classical Term	UED Equivalent	Meaning
Λ	$\nabla\sigma$	Entropy divergence, not cosmic push
G	ΔATT	Field misalignment response sensitivity
Force	$\delta E \cdot \Delta ATT$	Alignment-induced field effort
Curvature	∇R	Ripple memory gradient

Closing Note:

"In ripple mechanics, there are no constants—only consequences." UED abandons the belief that universal truths are fixed numbers. Instead, it recognizes that all observable dynamics arise from field-based, local re-alignment behaviors—measurable, inferable, and testable. Gravity and dark energy are not forces to be found. They are illusions born of ripple fatigue and observational echo.

Appendix B: Glossary of Ripple Terms for Physicists

Condensed UED vocabulary tailored for cosmology and field research.

This glossary is designed to help physicists, cosmologists, and field researchers transition from legacy scientific terms to the UED (Universal Energy Dynamics) lexicon. Each entry includes a concise definition, contrast with conventional terminology, and its relevance in ripple-based understanding.

ATT – Alignment Threshold Time

- **Definition:** The minimum time required for any energy system (field, particle, body) to realign with the local ripple field.
- **Classical Analog:** Reaction time, inertial delay, time dilation (loosely).
- **UED Insight:** Governs apparent motion, orbital stability, and entropy-based delays. ATT is not constant—it varies with local field stress and ripple memory.

RDI – Ripple Delay Index

- **Definition:** A scalar measure of how much delay a signal (like light or wave) experiences due to ripple fatigue or alignment disturbance.
- **Classical Analog:** Redshift, refractive index (loosely).
- **UED Insight:** RDI decouples observed delays from velocity and mass assumptions. High RDI zones can mimic

recessional motion without any movement.

SVE – Structured Vacuum Energy

- **Definition:** The pre-energetic, still-field fabric from which all ripples arise.
- **Classical Analog:** Vacuum field, aether (discredited), zero-point energy.
- **UED Insight:** SVE is not passive; it's structured, memory-bearing, and alignment-sensitive. It's the *medium of causality*, not a background.

Ripple Tension

- **Definition:** The local stress within the ripple field caused by misalignment or entropy divergence.
- **Classical Analog:** Gravitational potential, energy density (partial overlap).
- **UED Insight:** Where classical science sees "force," UED sees ripple tension—a directional field pressure that seeks equilibrium.

Ripple Memory

- **Definition:** The field's residual alignment history with past energy events.
- **Classical Analog:** None direct; closest might be gravitational waves or field remnants.
- **UED Insight:** Explains coherence in systems without ongoing interaction (e.g., entanglement, stable galaxy edges). Memory enables pre-ripple configuration prediction.

Entropy Gradient / σ

- **Definition:** A measure of how much alignment tension is lost or redistributed over space-time.
- **Classical Analog:** Entropy in thermodynamics, cosmological arrow of time.
- **UED Insight:** Drives relaxation, void formation, and directionality of ripple spread. Not just disorder—but *alignment exhaustion.*

Ripple Echo

- **Definition:** A secondary or delayed field response resulting from a prior alignment event.
- **Classical Analog:** Aftershock, radiation echo, or cosmological microwave background (CMB).
- **UED Insight:** Used to map past ripple disturbances and reconstruct the field's memory without assuming new mass or force.

Field Realignment

- **Definition:** The process by which a distorted field seeks equilibrium by modulating ripple tension.
- **Classical Analog:** Force equalization, harmonic oscillation.
- **UED Insight:** This is how motion occurs in UED—not as a push or pull, but as an *energy reallocation within a memory-bearing field.*

Ripple Fatigue

- **Definition:** Loss of ripple coherence due to prolonged tension or misalignment.
- **Classical Analog:** Energy dispersion, decoherence.
- **UED Insight:** Explains redshift anomalies, delayed alignment, and apparent cosmic "stretch" without invoking expansion or acceleration.

Realignment Gradient (∇◈)

- **Definition:** The vector measure of alignment pressure across regions of the ripple field.
- **Classical Analog:** None—unique to UED.
- **UED Insight:** Predicts ripple behavior, field flow, and the direction of entropy relaxation. Critical in spacecraft trajectory and large-scale structure analysis.

Ripple Field Topology

- **Definition:** The geometric and energetic map of ripple nodes, memory zones, and tension lines.
- **Classical Analog:** None directly; topological field theory is too abstract.
- **UED Insight:** Used to guide predictions in quantum systems, cosmic voids, and phase interference across space.

Entropic Delay

- **Definition:** The lag between an input alignment event and its full field response due to entropy saturation.
- **Classical Analog:** Signal lag, inertial delay, none directly.
- **UED Insight:** Explains time-lagged lensing, delayed decay, and supernova light curve variations.

Ripple Sink

- **Definition:** A region of the field where ripple alignment has been exhausted and tension diffused.
- **Classical Analog:** Gravitational well or entropy well (partial analogy).
- **UED Insight:** Cosmic voids are not empty—they're ripple sinks. Low ATT variance confirms exhaustion, not absence.

ATT Fluctuation Zone (AFZ)

- **Definition:** A space where ATT varies rapidly due to entropy turbulence or ripple memory conflict.
- **UED Insight:** Predicts danger zones for spacecraft misalignment, failed missions, or unexpected drag anomalies.

Phase Memory Shift

- **Definition:** A temporal offset in ripple response due to misaligned historical events embedded in field memory.
- **Classical Analog:** None directly.
- **UED Insight:** Gives rise to lensing without mass and non-causal entanglement patterns.

Closing Thought:

"The vocabulary of physics must evolve—not just to label new things, but to stop mislabeling old illusions."

These terms mark the transition from object-force dogma to field-alignment fluency. UED doesn't require us to abandon physics—it asks us to graduate it.

Appendix C: Observational Anomalies Re-explained

Galaxy rotations, lensing anomalies, cosmic flows—all reinterpreted through the ripple-based lens of Universal Energy Dynamics (UED).

This appendix revisits the most persistent anomalies in observational astrophysics—those traditionally attributed to "dark" causes—and reframes them using ripple alignment logic, without invoking hidden particles or exotic forces. Each case demonstrates how the UED framework restores consequence-based coherence to seemingly paradoxical data.

1. Flat Galaxy Rotation Curves

Conventional View: Outer stars in galaxies rotate at similar velocities to inner stars—defying Newtonian expectations, leading to the postulation of dark matter halos.

UED Reframe:

- These stars move within pre-aligned ripple tension zones, where memory fields preserve motion without needing extra mass.
- Ripple memory holds the outer zones in dragless harmonic balance.
- The field's energy configuration maintains orbital behavior—not local gravitational pull.

Prediction:

- Galaxies with violent merger histories or broken ripple

memory will show sharper falloff in rotation curves.

- Ripple delay index (RDI) and ATT mapping can reconstruct the ripple memory architecture.

2. Lensing Without Mass (Dark Lensing)

Conventional View: Gravitational lensing appears even where no visible or inferred mass exists, requiring unseen matter to warp spacetime.

UED Reframe:

- The lensing is caused not by missing mass but by **ripple phase distortion**—leftover field memory from prior energetic events.
- The field acts like a fluid refracting light along gradients of ripple exhaustion or compression.

Prediction:

- Light passing through ripple sinks or fatigue zones will bend without any mass nearby.
- Lensing intensity can be correlated with entropy gradient—not gravitational potential.

3. Cosmic Flows and Bulk Motion

Conventional View: Galaxy clusters exhibit coherent large-scale motion not explainable by gravitational dynamics alone—suggesting exotic scaffolding or initial condition fine-tuning.

UED Reframe:

- Cosmic flows reflect ripple **relaxation currents**—directional redistributions of tension across the structured vacuum.
- These are not "motion" in space, but **alignment gradients**

that carry systems with them like boats on a tide.

Prediction:

- Bulk motion will align with ripple stress lines, not necessarily with mass density.
- Mapping ripple relaxation fields will allow long-range flow prediction.

4. Redshift Irregularities (e.g., Discordant Galaxies)

Conventional View: Redshift is treated strictly as a velocity indicator—interpreted as space expansion or Doppler effect.

UED Reframe:

- Redshift arises from **ripple fatigue**—light loses coherence when traveling through tension-heavy or entropy-saturated zones.
- Different paths through the universe introduce different RDI profiles, skewing redshift data.

Prediction:

- Galaxies of similar distance but different RDI paths will show discordant redshift.
- Redshift is **not a universal velocity signature**, but a memory-lag measurement.

5. Early Galaxy Formation (Massive High-z Galaxies)

Conventional View: Distant (high-z) galaxies appear too massive or evolved too early for standard cosmological timelines.

UED Reframe:

- These galaxies formed in **early alignment zones** of the

ripple field where entropy was low and SVE was coherent.
- They did not "mature fast"—they emerged **fully formed** from structured field configurations.

Prediction:

- Early galaxy clusters should correlate with field symmetry, not accretion simulations.
- Starburst activity maps will show accelerated ignition where ripple relaxation preceded formation.

6. Cold Spot in CMB and Other Void Signatures

Conventional View: Interpreted as statistical anomalies, ISW effects, or relics of inflationary fluctuation.

UED Reframe:

- These are **zones of ripple silence**—regions where entropy has already relaxed, and the field no longer emits detectable response.
- Stillness itself distorts observational coherence—not lack of data, but lack of activity.

Prediction:

- Cold spots should coincide with regions of minimized ATT variance and RDI slope.
- Lensing or radiation skew should be absent or paradoxical in such zones.

7. Bullet Cluster and "Mass Displacement" Events

Conventional View: Colliding galaxy clusters show separation of visible matter and gravitational lensing zones—used as smoking-gun for dark matter.

UED Reframe:

- Lensing zones reflect **ripple phase memory**, not present mass.
- The field retains tension gradients from past mass alignments even after collision displaces the baryonic matter.

Prediction:

- Similar "displacement" events will be found where no collision occurred—just ripple misalignment.
- ATT disturbance zones will precede lensing shift, measurable before visible mass adjusts.

8. Large Quasar Groups (LQGs)

Conventional View: Presence of ultra-large quasar alignments violates cosmic isotropy and the scale of homogeneity.

UED Reframe:

- LQGs reflect long-chain **ripple alignment paths**, not physical structures.
- These are "strings of ignition" where memory traces allowed coherence over vast distances.

Prediction:

- No continuous matter bridges will be found between quasars.
- RDI and field topology maps will show alignment but not causally connected material.

Closing Thought:

"The anomalies weren't anomalies. They were answers—if only the questions had been different."

This appendix doesn't erase the observations—it realigns them with ripple-centric logic. Every contradiction under ΛCDM becomes a **consequence** under UED.

Appendix D: Detector Design for Future Science

B*lueprints for ripple-aware instruments based on ATT (Alignment Threshold Time) and RDI (Ripple Delay Index) sensitivity.*

Purpose of This Appendix

Current scientific instruments—from particle detectors to gravitational wave observatories—are designed to detect *events*, *forces*, or *particles*. However, under the Universal Energy Dynamics (UED) framework, these represent only *aftershocks* of ripple-field interactions—not the field itself. This appendix proposes a paradigm shift in instrumentation: from event-based sensors to *consequence-sensitive devices* that track field behavior, ripple fatigue, and entropic gradients.

1. Core Detection Philosophy: From Impact to Imprint

Legacy Detectors:

- Built to sense collisions, radiation, oscillations, or curvature.
- Operate by counting interactions (e.g., photons, neutrinos, gravitational waves).
- Miss the structured preconditions of those interactions.

Ripple-Aware Detectors:

- Aim to measure the *readiness* of a field to interact, not the interaction itself.
- Focus on alignment delay, field memory, and non-event disturbances.

- Use latency, harmonic tension, and entropy mapping to derive conclusions.

2. Primary Metrics for Detection

a. ATT — Alignment Threshold Time

- Measures the *time lag* between stimulus (ripple) and field readiness to respond.
- Indicates local entropy density and ripple stiffness.
- Can be used to predict successful alignment or failure in any mission (space, atomic, biological).

b. RDI — Ripple Delay Index

- Tracks how long it takes a ripple to propagate through a field segment relative to its expected path.
- Detects ripple fatigue, phase misalignment, and entropic decay.
- Especially useful in high-vacuum or deep-space zones.

c. ETR — Entropic Tension Response

- Proposed secondary metric for measuring distortion without interaction.
- Captures the energetic echo of alignment zones undergoing memory relaxation.

3. Conceptual Blueprints

a. Ripple Alignment Observatory (RAO)

- A ground- or space-based lab that maps ripple latency across large-scale environments.
- Instruments:
    - Interferometric delay arrays tuned to RDI fluctuations.
    - ATT-mapped chronometric fields (ultra-precise time grids).
    - MEMS-based ripple-tension discs to convert entropic tension into quantifiable variance.

b. Ripple Field Imager (RFI)

- Works like an MRI but for structured vacuum.
- Utilizes phase-locked coherent background sources (e.g., neutrino emitters or long-wave EM resonance).
- Detects:
    - Slow field swirls
    - Ripple depressions
    - Local alignment scars

c. ATT-Based Navigation Grid

- For spacecraft and deep-sea submersibles.
- Continuously measures ATT ahead of the vehicle to ensure optimal trajectory through energy-stable zones.
- Prevents "rip-inertia" or navigation loss due to unseen ripple turbulence.

4. Prototype Sensor Materials

- **Graphene-Tension Capacitors**: For fine measurement of micro-level alignment resistance.
- **Quantum-Coherent Fiber Lattices**: To register RDI delays in coherent phase bands.
- **Piezoelectric Vacuum Shells**: Sensitive to ripple pressure variance without requiring particle flux.

These materials don't register objects—they register *field coherence* and *alignment fatigue*.

5. Applications

Domain	Ripple-Aware Application
Space Navigation	ATT-mapped mission planning and reentry stability
Quantum Research	Identifying ripple-based entanglement zones without wave-particle paradox
Medical Imaging	Ripple imprints for early-stage tissue anomaly detection
Seismology	Measuring earth's ripple echo readiness to predict quakes before tremors
Fusion Research	Understanding ripple alignment zones for ignition predictability

6. Ripple Sensor Array Proposal (RSA-Net)

- A global or orbital network of synchronized ripple sensors.
- Continuously logs RDI/ATT variance across:
  - Atmosphere
  - Near-Earth field
  - Interplanetary voids
- Shared as open-source ripple maps to assist all fields—navigation, communication, cosmology, climate,

and even social behavior analytics (crowd dynamics through ripple sync).

7. Measurement Philosophy Shift

Old Paradigm	UED-Informed Paradigm
Count particles	Trace ripple phase lag
Observe force	Detect field fatigue and re-alignment resistance
Look for noise	Listen for silence and stillness zones
Model with constants	Model with real-time ripple dynamics
Define space	Map ripple memory and topology

Closing Insight:

"Detection doesn't begin when an event occurs. It begins when a field decides to respond."

Ripple-aware instrumentation doesn't *measure* reality—it *aligns* with it. These devices don't just record events; they *anticipate outcomes* by listening to the quiet signals that conventional physics overlooked.

Further Reading & References

Includes UED white papers, comparative works, and source critiques.

Purpose of This Appendix

This curated list provides foundational and supplemental material for readers seeking to deepen their understanding of Universal Energy Dynamics (UED), its philosophical foundations, scientific critiques, and the contrasting frameworks of legacy science. These readings are organized to guide physicists, philosophers, researchers, and open-minded learners from current limitations toward ripple-aware insight.

1. Foundational UED Documents (by Sandeep Chavan)

These core texts establish the UED framework, propose alternatives to classical physics, and open avenues for empirical realignment:

- **"Structured Vacuum Energy: The Still Foundation of All Motion"** Lays out the concept of SVE as the base of reality—before time, motion, or matter.
- **"Ripple Mechanics: The New Language of Physical Reality"** Introduces ripple alignment, ATT, and RDI as measurable quantities.
- **"The Ripple Equation: Replacing Gravity with Realignment"** Derives gravitational effects from ripple field memory rather than attractive force.
- **"ATT and RDI: A Framework for Measuring Entropic**

Delay" Defines field-based time perception as a threshold, not a dimension.

- **"The Cosmos Remembers: Entropy, Memory, and Universal Consequence"** Introduces ripple memory and historical field fatigue as alternatives to expansion models.
- **"Chavan's Realism: Redefining Science through Consequence"** Philosophical treatise on replacing force-centric and object-based inquiry with ripple-aware thinking.

These are available through UED research portals and preprint archives (Zenodo, Gyrus Institute, etc.).

2. Comparative Legacy Science Works

To understand UED's contrast with mainstream thinking, the following texts represent the dominant models UED critiques:

- **Stephen Hawking,** ***A Brief History of Time*** Relevant for its depiction of cosmological constants, time flow, and black hole paradoxes.
- **Sean Carroll,** ***The Particle at the End of the Universe*** Describes the Higgs boson and metaphysical extensions of the Standard Model.
- **Max Tegmark,** ***Our Mathematical Universe*** Reveals the assumption of reality as mathematical abstraction—UED challenges this.
- **Roger Penrose,** ***Cycles of Time*** Offers cyclic cosmology but still trapped in entropy and geometry as primary dimensions.
- **Einstein's Original Papers (1905–1916)** Especially General Relativity and the cosmological constant—used as critique targets in UED.

3. Philosophical Background & Metaphysical Critique

UED's worldview borrows selectively from philosophical traditions that resonate with consequence-based reality:

- **Bertrand Russell, *The Problems of Philosophy*** For understanding empiricism and perception limits.
- **Immanuel Kant, *Critique of Pure Reason*** Important to contrast classical intuition with ripple alignment awareness.
- **David Bohm, *Wholeness and the Implicate Order*** A rare physicist who leaned toward a non-fragmented reality, similar to ripple dynamics.
- **Samkhya & Vedanta Upanishadic Texts (Interpretive Editions)** As foundational echoes of field-based existence, non-force causality, and consequence tracing.
- **Fritjof Capra, *The Tao of Physics*** Useful as a bridge between Eastern wisdom and quantum metaphors, though UED moves further.

4. Experimental Proposals and Reports

Select reading for empirical thinkers interested in practical validation of UED predictions:

- **"ATT Variance in Low-Gravity Conditions"** Proposal for ripple re-entry navigation via ATT sensors (replacing orbital mechanics).
- **"Ripple Delay Index in Cosmic Void Lensing"** Report on field-mapped lensing without inferred mass presence.
- **"Field Realignment in Early Galaxies"** Observational reinterpretation of high-z galaxies via ripple memory, not inflation.

- **"Entropy Echo Mapping in Ripple-Thick Zones"** Designs for ripple field fatigue detectors and entropy latency monitors.
- **"Reframing Redshift: Ripple Fatigue Over Velocity"** Suggests new calibration methods for interpreting anomalous redshifts.

5. Suggested Reading Path

To help readers from diverse backgrounds:

Audience	Start With	Then Explore
General Thinker	*Chavan's Realism, A Brief History of Time* (as contrast)	*The Cosmos Remembers, Wholeness and the Implicate Order*
Physics Student	*Ripple Mechanics,* Einstein's 1915 Paper	*ATT and RDI, Cycles of Time*
Philosopher	*Problems of Philosophy, Critique of Pure Reason*	*Structured Vacuum Energy, The Tao of Physics*
Experimentalist	*ATT Variance in Low-Gravity Conditions*	*Entropy Echo Mapping, RDI in Cosmic Voids*

Closing Thought

"Science doesn't grow by adding more data. It grows when it subtracts its own illusions."

— Sandeep Chavan, UED Founder

If we continue searching for answers with the same tools that created the questions, we remain trapped. These readings offer a way out—a return not to fantasy, but to fundamental alignment.

Acknowledgments

For the scientists, observers, and the quiet field that never asked to be misunderstood.

This book—and the Universal Energy Dynamics (UED) framework it elaborates—is not the outcome of rebellion against science. It is the result of reverence for it.

I owe profound gratitude to the countless scientists, physicists, philosophers, and thinkers—past and present—who dared to peer into the unknown, even if the language they used was shaped by their times. To Newton, for teaching us how to measure; to Einstein, for teaching us how to bend; and to the many nameless observers who cataloged the cosmos long before they understood it—thank you for drawing the first ripples, even when you called them forces, particles, or equations.

To the developers of instruments, from Galileo's telescope to the Large Hadron Collider, who tried—however imperfectly—to bring the invisible into view: your pursuit, even when flawed, was beautiful. This work does not dismiss your contributions. It stands upon them—and then invites the next step.

To those who resisted the comfort of consensus and dared to say "something is off"—this book is written with you in mind. You may have been dismissed as fringe, idealistic, or even mystical. But the courage to ask "what if we're wrong?" is the only reason progress ever happens.

To the structured vacuum—the still field beneath all motion, which never once asked to be understood in human terms, but shaped reality with silent precision—thank you. If this book has offered any clarity, it is because you allowed it, not because we earned it.

To my readers, fellow researchers, and quiet seekers: may your curiosity never be colonized. May your tools always serve wonder, not replace it.

And finally, to the misalignments, blind spots, and paradoxes of legacy science—you were not mistakes. You were invitations. Without them, there would be nothing to realign.

— *Sandeep Chavan*

About the Author

Sandeep Chavan is an independent physicist, educator, and author who has quietly—but persistently—challenged the very foundations of how we perceive the universe. With a mind equally fluent in scientific reasoning and philosophical inquiry, Sandeep has written over 100 books and research papers spanning physics, metaphysics, human psychology, cosmology, and educational reform.

As the founder of *Gyrus Institute*, he has spent years teaching students not just how to solve equations, but how to think beyond them. His teaching style is grounded in first principles, deep curiosity, and the belief that true understanding comes not from memorization—but from alignment with reality itself.

Sandeep's flagship contribution to modern thought is **Universal Energy Dynamics (UED)**—a framework that redefines the nature of reality through ripple fields, structured vacuum energy, and consequence-based interaction. His work replaces outdated constructs like gravitational force, dark matter, and spacetime curvature with a ripple-aware view of the cosmos where motion, memory, and misalignment are the real actors.

Unbound by institutional constraints and unseduced by academic prestige, Sandeep works from the ground up—literally from tea shops and classrooms—where insight meets simplicity. His ideas have reached thinkers across disciplines, and his frameworks are beginning to spark serious interest among those disillusioned by the cracks in legacy models.

Whether decoding the symbolism of ancient philosophies or questioning Einstein's field equations with mathematical precision, Sandeep's work consistently points to a single message:

Reality was never fragmented. We just failed to ripple in tune with it.

This book is not just his theory. It is his lived philosophy. A culmination of decades spent asking one simple but dangerous question: *"What if we've misunderstood everything by looking too hard in the wrong direction?"*

If you've reached this far, you are not just a reader. You are now part of the ripple.

Don't miss out!

Visit the website below and you can sign up to receive emails whenever Sandeep Chavan publishes a new book. There's no charge and no obligation.

https://books2read.com/r/B-A-EPGPC-URERG

BOOKS 2 READ

Connecting independent readers to independent writers.

Did you love *Dark ≠ Hidden*? Then you should read *$E \neq mc^2$: The Collapse of Classical Assumptions*[1] by Sandeep Chavan!

[2]

What if everything we thought we knew about energy is incomplete?

For over a century, $\mathbf{E = mc^2}$ has symbolized the elegance of physics—a compact equation that married mass with energy and reshaped the modern scientific worldview. But what if it also marked the beginning of a deeper misunderstanding?

In *$E \neq mc^2$: The Collapse of Classical Assumptions*, independent physicist and philosopher **Sandeep Chavan** presents a provocative new vision of reality—one where energy isn't stored in mass, light isn't emitted from particles, and explosions are not releases, but realignments.

1. https://books2read.com/u/4DaMLe

2. https://books2read.com/u/4DaMLe

At the heart of this transformation is **Chavan's Realignment Equation**, a field-centric model that replaces classical concepts of mass, force, and emission with ripple-based coherence, structured vacuum dynamics, and a new limit on perception: the **Alignment Threshold Time** (ATT). What Einstein treated as constants—mass, light speed, time—are reinterpreted as consequences of ripple synchronization within an intelligent field.

This book isn't just theoretical—it rewrites the practical language of energy, light, and decay. From **fire and nuclear bombs** to **the Big Bang** and **black holes**, each phenomenon is re-explored through the lens of **Universal Energy Dynamics** (UED). Along the way, it challenges the foundations of particle physics, space-time models, and object-based reality.

Designed for scientists, engineers, philosophers, and curious thinkers alike, $E \neq mc^2$ blends rigorous insight with radical clarity. It features:

A full breakdown of the **Realignment Equation**Field-based alternatives to mass, inertia, and forceThe **Collapse Law**: why things vanish, not explodeImplications for measurement, engineering, and consciousnessA bold rethinking of **Einstein's legacy**—not discarded, but upgraded

This is not science fiction. It is a **post-Einstein doctrine** built on consequence, perception, and the hidden geometry of energy.

Whether you're a physicist looking for a new frontier, a student questioning the standard model, or a philosopher seeking a bridge between matter and meaning, this book opens a door into a universe that doesn't move—it **realigns**.

Nothing glows. Everything echoes. Nothing moves. Everything adjusts.

It's time to collapse old assumptions—and perceive energy as it truly is.

Read more at https://www.sandeepjchavan.com/.

Also by Sandeep Chavan

Chavan's UED research

CMGED Research Report

Chavan's Cosmic Continuum (C^3)

Frequency

Ripples of Reality

Reality Realigned

The Great Realignment

I – The Cosmic Leverage

Chavan's Cosmic Theory of Relativity (CTR)

Chavan's Theory of Light

Feeling Like a Physicist

TIME – The New Definition

$E \neq mc^2$: The Collapse of Classical Assumptions

The Chavan Codex

Chavan's Atomic Model

Chavan's Theory of Everything

Echoes of War: The Russia-Ukraine Conflict and Its Global Ripples

Shattered Lives

The War Economy

Borders Redrawn

Arms and Allies
Collateral Damage
Justice in the Ashes

Somewhere in the Darkness of the Cosmos
The Blueprint of the Universe
The Cosmic Breath
From the Edge of the Universe

The Eternal Sage Chronicles
The Sage and the Sovereign
The Family Pillar
The People's Flame
The Warrior's Soul
The Bridging Mentor
The Guiding Light

The Exodus
The Last Cycle: 2075

The Fundamental Theory of Life & Existence
Energy in Action: The Soul Demystified
The Active Mind
Energy in Connection
Death: The Disconnection
Subconsciousness: The Chosen Disconnection

Desires: The Engine of Life

The Restoration of Human Knowledge

Restoring Bhagavad Gita

The War of Thoughts

Cracks in the Known

The Human Game

The Divine Dilemma

The Original Intent

The Answers

Ultra Modern Life Philosophy

The Duality of Laughter and Tears

Take It Easy

Compare to Accept

Standalone

It's Not AI, It's AHI - Amplified Human Intelligence

The Decision Paradox

The Four Sapiens

Malicious Script of Indian Polity

The IIT Legacy & Global Impact

The IIT Legacy & Global Impact

You are just a Version of Your Original

Win the Game You Didn't Choose

Always Keep Your Bags Packed
Reveal to Shield - Learn the Game of Facades
Engage Beyond Elections
Culture, Identity & Change: The Evolution of Indian Society
The Triangular Dynamics
Never Trust Without Doubt
If You Can't Change the Country, Leave the Country
Live Fearless with Fear
India: The Investment Magnet
Realignment
Powering a Billion Dreams
Universal Skill
Extension: From Brush to Algorithm
Human Side of Einstein
The Future Knows No Lines
Let's Bring the Generations to the Table
Unified Energy Dynamics
A Beat That Never Dies
Strategic Horizons
Right From Scratch
The Forgotten Innovators
Physics Fixing
The Savior Illusion
The Selfish Side
The Age of the Weakest Sapiens
Neural Wars
The Absorption: There Is No Death
Strange Therapist
Bubbles: Become a Drop
Beyond the Blue
Half-Scholar
The Law of Truth
Failure

Philosophy 2.0: A Structural Manifesto
Chaos: A Gentle Guide to the Unresolvable
Self-Help: New Syllabus
Rippling
Game: How to Play the Share Market
Out of Syllabus
Saffronized: The Code Behind the Crown
The Anger
Moksha: The Liberation
Chavan vs Einstein: The Universe Was Never Curved
Rocket Science: The Ripple Gambling
The Cosmos: The Still Universe
The Taste of Silence
Chavan's Post-Materialist Human Model
10-Second Rule
Chavan vs Hawking
It was Never a Wave Signal
Quantum Courtroom: The Trial of a Century
Chavan–Heisenberg Ripple Uncertainty Principle
The Great Disruption
Dark ≠ Hidden

Watch for more at https://www.sandeepjchavan.com/.